Léon Delbos

Nautical Terms

In English and French with Useful Tables

Léon Delbos

Nautical Terms
In English and French with Useful Tables

ISBN/EAN: 9783337396985

Printed in Europe, USA, Canada, Australia, Japan

Cover: Foto ©berggeist007 / pixelio.de

More available books at **www.hansebooks.com**

NAUTICAL TERMS

In English and in French.

WITH USEFUL TABLES.

BY

LEON DELBOS, M.A.

H.M.S. BRITANNIA.

WILLIAMS AND NORGATE,

14, HENRIETTA STREET, COVENT GARDEN, LONDON;
AND 20, SOUTH FREDERICK STREET, EDINBURGH.

1889.

PREFACE.

THIS Vocabulary of Nautical Terms possesses at least one merit, that of accuracy. The manner in which it has been compiled can scarcely leave any doubt on that head. All the terms referring to the parts of the ship, or to her rigging —in short, all words used in seamanship—have been verified by means of illustrated and explanatory diagrams in each language. Besides this, some of the best books have been consulted, and especially Captain Nares' 'Seamanship,' and the most excellent translation of it by Lieut. Ed. Tiret of the French Navy.

The words of command have been tested over and over again—a very necessary thing, considering the difficulty of this portion of the work.

The utility of such a book is, I think, obvious. Officers of the Merchant Service, who have such frequent occasions of ordering repairs whilst in a foreign port, or of employing either French or French-speaking pilots, will find it no less useful than officers of the Royal Navy.

LEON DELBOS.

H.M.S. BRITANNIA,
DARTMOUTH,
January, 1889.

CONTENTS.

INTRODUCTION.

SINCE this Vocabulary of Nautical Terms appeared, at the beginning of the present year, it has undergone constant revision. The Compiler has found, by practical use, that very important words, used especially in modern naval warfare, had been omitted, owing to the difficulty of collecting together all the useful words without burdening the book with long columns of obsolete terms. These words have now been added before the entire edition is exhausted, so that present purchasers may have at once the benefit of them. We have also taken this opportunity of making a few necessary corrections.

LEON DELBOS.

H.M.S. BRITANNIA, DARTMOUTH,
May, 1889.

PAGE

1—aback (to be taken).   faire chapelle.

2—aft (haul aft jib sheet !).   hâlez le foc !

2—altitude (meridian).   la hauteur méridienne.

3—armour (of ships).   blindage, *m.*, cuirasse, *f.*

3—aye, aye ! (*answer to an order*).   bien monsieur! bien capitaine !

answer to a challenge:—*The coxswain, in the French Navy, answers as follows: If there is in the boat an officer superior to the rank of captain*—Amiral ! *If the officer is the captain of the ship*—Commandant ! *If he is the captain of another ship, the name of the ship is mentioned first*—'Colbert,' commandant ! *If any other superior officer*—Officier supérieur ! *If any other officer*—Officier ! *If there is no officer, and the boat is going to the ship which challenged her, the coxswain answers*—A bord ! English 'No, no !' *If not going on board that ship*—Rade !

4—balcony (stern walk).   la galerie de poupe.

4—belt (of armour).   une ceinture blindée.

5—berth (to give a wide —). éviter.
6—board (to) (as an enemy). aborder, &c.
6— „ „ (as a friend). aller à bord.
6—block (mooring). une ancre de corps mort.
7—bore (of a gun). calibre, *m.*, âme, *f.*
8—bowsprit (to run in the —). rentrer le beaupré.
8—brace (to) the yards. brasser les vergues.
9—buffers (hydraulic). frein hydraulique, *m.*
10—bunk. couchette, *f.*
10—bunker (coal). la soute au charbon.
10—cadet (naval). un élève de marine.
11—capstan (steam). cabestan à vapeur.
11—cashier (to). casser.
12—channel (buoyed —). une passe balisée.
13—coffer dam. bâtardeau, *m.*
14—column (the rear —). la colonne de queue.
15—compass error. la variation du compas.
15—conning tower. le kiosque de la barre.
15—course (to alter the —). changer la route.
15— *to steer a —.* *suivre une route donnée.*
16—cross (to) (like 2 ships). se croiser.
16—cruiser. un croiseur.
16— *torpedo —.* [east. *un croiseur torpilleur.*
16—current (the current sets le courant porte à l'est.
17—derrick. un mât de charge.
17—deviation (of compass). la déviation.
17—direction. direction, *f.*, gisement, *m.*
17— *directions.* *instructions, f. pl.*
17— *sailing —s for the* *instructions sur la Manche.*
        English Channel.
18—draft (what is your — of Combien calez-vous ?
        water ?)
21—float (to — a ship). mettre un navire à flot.
21—flood. le flux.
21— — *tide.* *marée, f., flux, m.*

21—fly (to let —).	larguer en bande.
21—foggy.	brumeux.
21—fore.	avant, de l'avant.
22— — *rigging.*	*le gréement de l'avant.*
22—forge (to) ahead.	courir de l'avant.
23—garnet.	un palan de charge.
23—gear.	les apparaux, *m.*
23— *to throw into —.*	*enclancher.* [*l'électricité.*
23— *electric firing —.*	*dispositif de mise à feu par*
23— *steam steering —.*	*barre à vapeur, f.*
23— *turret turning —.*	*le vireur de la tourelle.*
23— *training —.*	*les engins de pointage, m.*
24—guard ship.	le ponton amiral.
24—gun.	canon, *m.*
24— *to scale a —.*	*flamber un canon.*
24— *machine —.*	*mitrailleuse, f.*
24— *quick firing —.*	*canon à tir rapide.*
24— *saluting —s.*	*feux de salut.*
24— — *mountings.*	*les garnitures d'un canon.*
24—gunner.	canonnier, *m.*
24—hail (to) a boat.	héler.
24—halliard (peak).	martinet, *m.*
25—haul (to) up.	remonter, carguer.
26—heave (to) short.	virer à pic.
26—helm a weather !	la barre au vent !
26—put the helm down !	la barre dessous !
27—hour-glass.	ampoulette, *f.*
27—house (to) a mast.	caler un mât.
28—impulse (*for projection of torpedo*).	l'impulsion.
28—irons (fetters).	les fers, *m. pl.*
28— *in — (ship).*	*faire chapelle.*
28—jib (down jib !)	hâlez bas le grand foc !
28— *ease off the jib sheets !*	*filez les focs !*

PAGE	
31—light ship.	un bateau feu.
31— *search light.*	*projecteur, m.*
32—line abreast.	la ligne de file.
32— *— ahead.*	*la ligne de front.*
32— *quarter —.*	*la ligne de relèvement.*
32—luff a lee!	loffe toute!
32—lug sail (to dip a)	gambéyer.
33—mast-head (to be at the).	être en vigie.
33— *— man.*	*l'homme de vigie.*
35—moorings (at her —).	au mouillage.
35—mooring swivel.	un émerillon d'affourche.
35—no, no!	à bord! see p. 3, under *aye, aye!*
36—off! (shove).	poussez!
36—paul (to) the capstan.	mettre les linguets au cabstan.
37—paymaster.	commissaire, *m.*
37— *assistant —.*	*aide-commissaire.*
37—put (to) into D.	relâcher à D.
38—propeller.	propulseur, *m.*
39—ram (to).	aborder à l'éperon.
39— *hydraulic —.*	*un piston hydraulique.*
39—range (of weapons).	la portée.
39—recess.	encastrement, *m.*
39—rifle (magazine —).	fusil à répétition.
40—rocket.	fusée, *f.*
40—rope (tow).	remorque, *f.*, faux-bras, *m.*
41—sail (to hoist a —)	hisser une voile.
42—screws (twin).	hélices jumelles, *f.*
43—search light.	projecteur, *m.* [der *to veer.*
44—shackle.	maillon, *m.*, manille, *f.* : see un-
45—shield (for guns).	masque d'acier, *m.*
45—shove off!	poussez!
45—sound (to).	sonder.
45—soundings (to take the).	le sondage.
46—speed.	vitesse, *f.*

PAGE

46—speed. *full — ahead!*	*en avant, toute vitesse!*
46— *full — astern!*	*en arrière, toute vitesse!*
46—sponge (for gun).	écouvillon, *m.*
47—stay (long).	virer à long piç.
47—steam (to reverse).	renverser la marche.
47— *— steering gear.*	*une barre à vapeur.*
47— *—trials (of an engine).*	*faire les essais d'une machine.*
48—steerage way.	l'erre, *m.*
48—stern walk (balcony).	la galerie de poupe.
48—stern way (to make).	aller en culant.
49—stranded (to be).	s'échouer.
49—swab (to) the deck.	fauberder le pont.
50—tack.	amure, *f.*
50— *to be on the port —.*	*être bâbord amures.*
50— *to be on the starboard—.*	*être tribord amures.*
50—tackle (purchase).	palan, *m.*, caliorne, *f.*
50—tender (to a vessel).	une navire annexe.
51—tide (it is high —).	la mer est haute.
51— *it is low —.*	*la mer est basse.*
51—tiller.	la barre.
51— *— ropes.*	*drosse, f.*
51—tops (military).	hunes (*f.*) de combat.
52—torpedo booms.	espars à torpilles.
52— *— gun.*	*un tube de lancement.*
52— *— net.*	*un filet à torpille.*
52— *— port.*	*un sabord de lancement.*
52— *— warfare.*	*le maniement des torpilles.*
52— *to discharge a —.*	*lancer une torpille.*
52—towing hawser.	remorque, *f.*, faux bras, *m.*
53—unship oars! [pass).	désarmez vos avirons!
53—variation (of the com-	la déclinaison.
53—veer (to) 3 shackles.*	filer 3 maillons.

---

* The English shackle is equal to 12½ fathoms, and the French *maillon* to 30 mètres, or very nearly 16½ English fathoms.

NAUTICAL TERMS

IN ENGLISH AND IN FRENCH.

———◆———

aback (to be).	être masqué.
throw all — !	*masquez partout.*
heave all — !	„ „
sails —.	*en panne.*
abaft.	à l'arrière.
able-bodied seaman.	gabier breveté.
aboard.	à bord.
haul the tacks — !	*amurez !*
all — !	*embarquez !*
about (to put).	louvoyer.
accommodation.	aménagements, *m. pl.*
— ladder.	*échelle de commandement.*
accost, *v.*	accoster, aborder.
admiral.	amiral, *m.*
vice —.	*vice-amiral.*
rear —.	*contre amiral.*
admiralty.	amirauté, *f.* In France : le ministère de la marine.
adrift.	en dérive.
to get — upon a lee shore.	*s'affaler à la côte.*
adze.	herminette, *f.*
afloat.	à flot, sur mer.
aft.	arrière, *m.*
the wind is dead —.	*le vent est entre deux écoutes.*
the wind is right —.	*le vent est droit derrière.*

B

haul aft jib!	*hâlez le foc!*
haul — the fore sheet!	*hâlez l'écoute de misaine!*
aground.	échoué, à terre.
to run —.	*s'échouer.*
ahead.	en avant.
go —!	*en avant!*
ahoy!	oh!
ship —!	*ohé! du navire.*
air.	air, *m.*
— shaft.	*le puits d'aérage.*
alee.	dessous.
helm's —.	*la barre dessous.*
almanac.	almanach, *m.*
nautical —.	*la connaissance des temps.*
— — office.	*le bureau des longitudes.*
all right! all's well!	c'est bien! bon quart!
aloft.	en haut, dans la mature.
aloft (on the look-out).	*en vigie.*
away —!	*en haut les gabiers!*
alongside.	bord à bord.
to come —.	*accoster.*
aloof.	au large.
altitude.	hauteur, *f.*, élévation, *f.*
amidships.	par le travers.
helm —!	*droite la barre!*
anchor.	ancre, *f.*
the arms.	*les bras.*
the bill.	*le bec.*
the crown.	*le diamant.*
the flukes.	*les pattes, f.*
the ring.	*la cigale.*
the shank.	*la verge.*
the shoe.	*la sole, la sabatte.*
the stock or *beam.*	*le jas.*
bower anchor.	*ancre de bossoir.*

best bower anchor.	*la seconde ancre.*
kedge —.	*ancre à jet.*
sheet —.	*ancre de veille.*
stream —.	*ancre de touée.*
mooring block.	*corps mort, m.*
anchor (to).	ancrer, mouiller.
to cast —.	*jeter l'ancre.*
to cat the —.	*caponner l'ancre.*
to drag the —.	*chasser sur les ancres.*
to fish the —.	*traverser l'ancre.*
to heave up the —.	*déraper l'ancre.*
to ride at —.	*être à l'ancre.*
to weigh —.	*lever l'ancre.*
let go the — !	*mouillez !*
let go the starboard — !	*mouillez tribord !*
anchorage.	mouillage, *m.*
aneroid.	anéroide, *m.*
angle-iron.	cornière, *f.*
apron (timber).	contre-étrave, *f.*
armour.	blindage, *m.*
armourer.	armurier, *m.*
artillery.	artillerie, *f.*
heavy —.	*grosse artillerie, f.*
ashore.	à terre, sur terre.
astern.	de l'arrière, arrière.
to back —.	*nager à culer.*
— *of us.*	*derrière nous.*
go — !	*en arrière !*
athwart.	par le travers de, en travers de.
atrip (the anchor is).	dérapée (l'ancre est).
auxiliary screw (ship with).	un batiment mixte.
avast !	tenez bon !
awning.	tendelet, *m.*, tente, *f.*
axe.	hache, *f.*
aye, aye ! (answer to a call).	voila !

aye, aye! (answer to an order). bien capitaine !
— — (answer of coxswain). à bord !

back astern !	nage à culer !
	(on steamers) en arrière !
„	
to back a sail.	masquer une voile.
back stays.	galhaubans, *m. pl.*
flying —.	*galhaubans volants.*
backing and filling.	dériver vent dessus vent dedans.
balcony.	la galerie de poupe.
bale sling.	élingue (*f.*) en double.
bale out (to).	vider.
baler.	escope, *f.*
ballast.	lest, *m.*
to —.	*lester.*
to sail in —.	*aller sur lest.*
bar of the capstan.	une barre de cabestan.
— *of iron.*	*barre de fer.*
barge.	allège, *f.*, gabare, *f.*
— (*admiral's*).	*canot de l'amiral.*
bargeman.	marinier, *m.*, batelier, *m.*
barque.	un trois-mats.
barometer.*	baromètre.
the — *is rising.*	*le baromètre monte.*
the — *is falling.*	*le baromètre descend.*
barrel (of a gun).	canon, *m.*
— (*cask*).	*baril, m.*
basin (of a dock).	bassin, *m.*
batten.	batte, *f.*
to — *down the hatches.*	*condamner les panneaux.*
beach.	plage, *f.*, grève, *f.*
to —.	*échouer.*

---

* For comparison between English and French barometric scales, see Table V. at the end of the book.

beacon.	balise, *f.*
beam (of deck).	bau, *m.*
breadth of —.	*largeur, f.*
on her — *ends.*	*sur le côté.*
Bear (the great).	la grande ourse.
to — *away.*	*porter au large.*
to — *down.*	*arriver sur.*
to — *up.*	*laisser arriver.*
bearing (to take the) of.	relever.
—*s (engineering).*	*coussinets, m. pl.*
— *(the).*	*le relèvement.*
beat to windward (to).	louvoyer.
to — *out to sea.*	*gagner le large.*
belay (to).	amarrer.
belaying cleat.	taquet, *m.*
— *pin.*	*cabillot, m.*
bell.	cloche, *f.*
below.	en bas.
bend.	nœud, *m.*
to — *a rope.*	*frapper.*
to — *(sails).*	*enverguer.*
to — *(jibs).*	*garnir.*
to — *(a cable).*	*étalinguer.*
bends.	les préceintes, *f.*
double —.	*nœud d'écoute double.*
fisherman's —.	*nœud d'étalingure de grelin.*
single —.	*nœud d'écoute.*
studding sail halliard —.	*nœud de drisse de bonnette.*
carrick —.	*nœud d'ajut, nœud de vache.*
berth (cot).	gîte, *m.*
— *(post).*	*poste, m.*
— *(alongside a wharf).*	*le poste d'amarrage.*
sick —.	*le poste des malades.*
bight (of a rope).	double, *m.*
bilboes (punishment).	les fers, *m. pl.*

bilge water.	eau de la cale.
— *pieces.*	*quilles latérales, f.*
bill of health.	la patente de santé.
— *of lading.*	*connaissement, m.*
— *board.*	*plan incliné de traversière.*
billows.	vagues, *f. pl.*
binnacle.	habitacle, *m.*
bitts.	bittes, *f. pl.*
main —.	*grandes bittes.*
block (pulley).	moufle, *m.*
snatch —.	*poulie coupée, f.*
— *maker.*	*poulieur, m.*
blockade.	blocus, *m.*
to —.	*bloquer.*
— *runner.*	*forceur de blocus.*
blow (to).	souffler.
the wind —*s from the west.*	*le vent souffle de l'ouest.*
it is —*ing hard.*	*il fait une tempête.*
it is —*ing fresh.*	*il fraichit.*
blue peter.	le pavillon de partance.
bluff (cliff).	falaise, *f.*
board (on).	à bord.
the hands on —.	*les hommes du bord.*
to make a stern —.	*culer.*
on — *my ship.*	*à mon bord.*
to —.	*aborder, monter à l'abordage, prendre à l'abordage.*
board (to).	aborder (as an enemy).
boat.	canot, *m.*, embarcation, *f.*
ferry —.	*bateau de passage.*
fishing —.	*bateau pêcheur.*
steam —.	*bateau à vapeur, m.*
sailing —.	*bateau à voiles, m.*
life —.	*canot de sauvetage, m.*
long —.	*chaloupe, f.*

jolly boat.	*petit canot, m.*
pleasure —.	*bateau d'agrément.*
open —.	*canot non ponté.*
— *hook.*	*gaffe, f.*
— *ahoy!*	*ohé du canot!*
boatman.	batelier, *m.*
boatswain.	maître d'équipage, *m.*
—'s *mate.*	*contre-maître, m.*
bobstay.	la sous barbe de beaupré.
body post.	étambot avant, *m.*
boiler.	chaudière, *f.*
— *plate.*	*tôle, f.*
— *tubes.*	*bouilleurs, m. pl.*
bollard heads.	têtes (*f.*) d'allonges.
bolster.	coussin (*m.*) d'élongis.
bolt.	cheville, *f.*
eye —.	*cheville à œillet, f.*
ring —.	*cheville à boucle, f.*
screw —.	*boulon taraudé, m.*
starting —.	*repoussoir à manche, m.*
— *rope.*	*ralingue, f.*
— *rope needle.*	*aiguille à ralingue, f.*
boom (of harbour).	chaine, *f.*, estacade, *f.*
— (*of a sail*).	*bout dehors, m.*
— *irons.*	*blins, m.*
jib —.	*bout dehors de foc, m.*
flying jib —.	*bout dehors de clin foc, m.*
spanker —.	*gui, m.*
studding-sail —.	*bout dehors de bonnettes.*
to ride by the swinging —.	*être au tangon.*
bore (of gun).	calibre, *m.*
— (*on tidal river*).	*mascaret, m., barre, f.*
bottom (of ship).	carène, *f.*
in British —*s.*	*sous pavillon anglais.*
in French —*s.*	*sous pavillon français.*

to sink to the bottom.	couler à fond.
double —.	double coque, f.
flat bottomed.	à fond plat.
copper —.	doublé en cuivre.
bottomry.	bomerie, f.
— bond.	un contrat à la grosse.
— interest.	profit maritime, m.
bound to, or for . . . (ship).	navire en partance pour. . . .
outward —.	en partance.
homeward —.	en retour.
where are you — to ?	où allez vous ?
bow.	l'avant, m., le bossoir.
on the port —.	à, or par babord devant.
to cross the —.	courir à bord droit.
bowline.	bouline, f.
— on the bite.	nœud de chaise double.
running —.	nœud d'anguille.
— bridles.	branches de bouline.
bowse (to).	palanquer.
bowsprit.	beaupré, m.
box the compass (to).	réciter la rose des vents.
boxhaul (to).	virer vent arrière, virer lof pour
boxing off.	contrebasser devant. [lof.
boy.	mousse, m.
brace.	bras, m.
to —.	brasser.
to — up.	orienter au plus près.
lee —.	le bras de dessous le vent.
weather —.	le bras du vent.
main —.	grands bras.
fore —.	bras de misaine.
main top sail —.	bras de grand hunier.
fore top sail —.	bras de petit hunier.
main top gallant —.	bras de grand perroquet.
fore top gallant —.	bras de petit perroquet.

mizzen top sail brace.	*bras de perroquet de fougue.*
cross jack —.	*bras barré.*
mizzen top gallant —.	*bras de la perruche.*
sprit sail —.	*bras de la contrecivadière.*
top sail —.	*bras de hune.*
top gallant —.	*bras de perroquet.*
brail.	cargue, *f.*
to — up.	*carguer.*
to bream.	chauffer.
breastwork (of a ship).	fronteau, *m.*
— (*of the poop*).	*fronteau de la dunette.*
breech (of a gun).	culasse, *f.* [*la culasse.*
— *loader.*	*canon,* or *fusil se chargeant par*
breeching.	brague, *f.*
breeze.	brise, *f.*
light —.	*petite brise,* f.
bridge.	pont, *m.*
— (*of steam-boat*).	*passerelle, f.*
swing —.	*pont tournant, m.*
bridles (of the bowline).	branches, *f. pl.* (de boulines).
brig.	brick, *m.*
brine-pump.	pompe d'exhaustion, *f.*
to bring to.	mettre en panne.
she is brought to.	*son erre est coupée.*
to broach to.	lancer dans le vent, embarder
broadside (of a ship).	côté, *m.* [au vent.
to fire a —.	*tirer une bordée..*
broker.	courtier, *m.*
ship —.	*courtier maritime, m.*
brokerage.	courtage, *m.*
bucket.	seau, *m.*
buckler.	gatte, *f.*
builder (ship).	constructeur, *m.*
bulk.	charge, *f.*
laden in —.	*en vrac.*

to break bulk.	commencer le déchargement.
bulkhead.	cloison, f.
bull's-eye.	cosse de bois, f.
bullet.	balle, f.
bulwarks (of ships).	pavois, m.
bumboat.	bateau à provisions.
bumkin.	minot, m.
bunker.	soute, f.
coal —.	soute au charbon.
bunt.	le sein d'une voile.
buntlines.	cargues-fonds.
buoy.	bouée, f.
life —.	bouée de sauvetage, f.
— rope.	orin, m.
to —.	baliser.
buoyancy.	légèreté, f.
buoyant.	flottant.
burden (of a ship).	contenance, f., port, m.
ship 3000 tons —.	vaisseau du port de 3000 ton-
burgee.	guidon, m. [neaux.
burton pendants.	pantoires de candelettes, f.
cabin.	chambre, f., cabine, f.
— (captain's).	chambre du commandant.
fore —.	cabine d'avant, f.
state —.	cabine de première classe, f.
— passenger.	passager de première classe, f.
— boy.	mousse, m.
cable.	cable, m.
—'s length.	encâblure, f.
caboose.	la cuisine.
cadet.	élève de marine, m.
calk.	calfater.
calker.	calfat, m.
calking.	calfatage, m.

calking iron.	*calfait, m.*
callipers.	compas de calibre, *m.*
calm.	calme, *m.*
dead —.	*calme plat.*
can-hook.	une élingue à pattes, *f.*, une
canister shot.	mitraille, *f.* [patte à futailles.
cannon.	canon, *m.*
— *ball.*	*boulet de canon, m.*
cap.	chouquet, *m.*
cape.	cap, *m.*
capshore.	épontille, *f.*, chouque, *f.*
capsize.	chavirer.
capstan.	cabestan, *m.*
to man the —.	*mettre du monde au cabestan.*
man — *!*	*au cabestan !*
to heave in at the —.	*virer au cabestan.*
captain (of a man-of-war).	capitaine de vaisseau, *m.*
— (*merchant navy*).	*capitaine au long cours, m.*
— (*of the watch*):	*chef de quart, m.*
— (*of the gun*).	*chef de pièce, m.*
careen (to).	abattre en carène.
careening-place.	carénage, *m.*
cargo.	cargaison, *f.*, chargement, *m.*
general —.	*un chargement en cueillette.*
carling.	traversin, *m.*
carpenter (of a ship).	charpentier, *m.*
chief —.	*le maître charpentier.*
cartridge (of rifle).	cartouche, *f.*
— (*of gun*).	*gargousse, f.*
casting.	faire son abattée.
cast loose (the ropes) (to).	larguer.
cat.	capon, *m.*
to — (*the anchor*).	*caponner l'ancre.*
— *block.*	*poulie (f.) de capon.*
— *head.*	*bossoir (m.) de capon.*

cat-o'-nine-tails.   garcette, f.
—'s-paw.   gueule (f.) de raie.
catch hold !   attrapez !
caulk.   see calk.
chain.   chaîne, f.
— plates.   cadènes, f.
— shot.   boulets ramés, m. pl.
— s.   porte-haubans, m. pl.
— wales.   porte-haubans, m. pl.
fore chains, &c.   see under fore and main.
challenge (a ship) (to).   héler.
channel.   canal, m., détroit, m.
— (of harbour).   passe, f.
—s (of rigging).   porte-haubans, m.
chapelling.   faire chapelle.
chaplain.   un aumonier.
chart.   carte marine, f.
charter (to).   frêter.
— (party).   la charte partie.
charterer.   frêteur, m.
a ship chartered by govern- un navire frêté par l'état.
check-piece.   [ment. safran, m.
checks.   jottereaux, m. pl.
chock (a).   une cale.
chopping sea.   une mer clapoteuse.
chronometer.   chronomètre, m.
to rate the —s.   régler les chronomètres.
circle.   cercle, m.
clear.   dégagé.
to —.   dégager.
to — a vessel.   passer un vaisseau.
to — up (of the weather). s'éclaircir.
cleat.   taquet, m.
clew.   point, m.
— garnets.   cargue-fonds, m.

to *clew up*.   *carguer*.
— *line*.   *cargue-point, m.*
cliff.   falaise, *f.*
clinch.   cable, *m.*, étalingure, *f.*
inside —.   *nœud de bouline double.*
outside —.   *nœud de bouline simple.*
clipper.   fin voilier, *m.*, clipper, *m.*
close-reefed.   au bas ris ; à la cape.
— *in shore*.   *près de terre.*
— *hauled*.   *au plus près.*
cloud.   nuage, *m.*
cloudless.   sans nuages.
cloudy.   nuageux, sombre.
it is —.   *le temps est couvert.*
it is getting —.   *le temps se couvre.*
coal.   charbon de terre, *m.*; houille, *f.*
to —.   *faire son charbon.*
— *dépôt*.   *dépôt de charbon.*
— *bunkers*.   *la soute au charbon.*
coamings.   hiloire, *f.*, surbaux, *m. pl.*
coast.   côte, *f.*, littoral, *m.* [*danger*.
the — *is clear*.   *la côte est belle*, or *il n'y a pas de*
to —.   *côtoyer.*
coaster.   caboteur, *m.*
coasting.   navigation (*f.*) côtière.
— *trade*.   *cabotage, m.*
cock-bill (to) the anchor.   faire peneau.
cockpit.   le poste de malades.
cockswain.   patron de chaloupe.
coil (of rope).   glène, *f.*
to —.   *lover.*
collier (ship).   charbonnier, *m.*
colours.   le pavillon national.
hoist the —!   *hissez le pavillon !*
strike the —!   *amenez le pavillon !*

column.	colonne, *f.*
the leading —.	*la colonne de tête.*
the starboard wing —.	*la colonne de droite.*
the port wing —.	*la colonne de gauche.*
the leader of a —.	*le vaisseau de tête,* or *chef de file.*
commander.	capitaine de frégate.
commission (ship in).	vaisseau (*m.*) en armement.
commodore.	chef d'escadre. [mer.
compass.	boussole, *f.*, compas (*m.*) de

NORTH.	NORD.
N. b. E.	N. ¼ N.E.
N.N.E.	N.N.E.
N.E. b. N.	N.E. ¼ N.
N.E.	N.E.
N.E. b. E.	N.E. ¼ E.
E.N.E.	E.N.E.
E. b. N.	E. ¼ N.E.
EAST.	EST.
E. b. S.	E. ¼ S.E.
E.S.E.	E.S.E.
S.E. b. E.	S.E. ¼ E.
S.E.	S.E.
S.E. b. S.	S.E. ¼ S.
S.S.E.	S.S.E.
S. b. E.	S. ¼ S.E.
SOUTH.	SUD.
S. b. W.	S. ¼ S.O.
S.S.W.	S.S.O.
S.W. b. S.	S.O. ¼ S.
S.W.	S.O.
S.W. b. W.	S.O. ¼ O.
W.S.W.	O.S.O.
W. b. S.	O. ¼ S.O.

WEST.	OUEST.
W. b. N.	O. ¼ N.O.
W.N.W.	O.N.O.
N.W. b. W.	N.O. ¼ O.
N.W.	N.O.
N.W. b. N.	N.O. ¼ N.
N.N.W.	N.N.O.
N. b. W.	N. ¼ N.O.

N.B.—On the French compass, ¼ is pronounced *quart*, somewhat like *car* in *carmine*.

compass.	
azimuth —.	*compas de variations.*
a point of the —.	*une aire de vent.*
standard —.	*le compas étalon.*
steering —.	*le compas de route.*
— *course.*	*la route au compas.*
— *error.*	*correction du compas.*
a pair of — *es.*	*un compas, m.*
compressor.	étrangloir, *m.* [*m.*
condenser.	condenseur, *m.*, condensateur,
condensing engine.	une machine à condensation.
„	machine à basse pression.
conning.	faire gouverner.
consignee.	consignataire, *m.*
consul.	consul, *m.*
consulate.	consulat, *m.*
convoy.	convoi, *m.*
cook.	coq, *m.*
copper.	cuivre, *m.*
— *bottomed.*	*doublé en cuivre.*
cot.	cadre, *m.*
counter.	voûte, *f.*
course (of a ship).	la route.
fore —.	*la misaine.*

main course.	*la grand'voile.*
courses.	les basses voiles.
coxswain.	un patron de chaloupe.
cradle.	ber, *m.*
craft.	embarcation, *f.*
crane.	grue, *f.*
steam —.	*grue à vapeur.*
crank.	manivelle, *f.*
— (*a ship that is*).	*vaisseau portant mal la toile.*
creek.	anse, *f.*, crique, *f.*
crew.	équipage, *m.*
gun's —.	*servants, m. pl.*
cringle.	patte, *f.*
cross-sails.	*see* sails.
— *trees.*	*barres traversières, f. pl.*
crossing (by sea).	traversée, *f.*
crow-bar.	pince, *f.*
crow's-foot.	une araignée de tente.
crowd all sail (to).	forcé de voiles.
cruise (to).	aller en croisière, croiser.
cruise.	croisière, *f.*, campagne, *j.*
crutches.	dames, *f.*
current.	courant, *m.*
custom-house.	douane, *f.*
— *officer.*	*douanier, m.*
to pass the customs.	*passer en douane.*
— *duty.*	*droit de douane.*
cutter.	côtre, *m.*
1st —.	*canot numéro 1.*
2nd —.	*canot numéro 2.*
cutwater.	taille-mer, *m.*
cylinder.	cylindre, *m.*
davits.	les bossoirs d'embarcation, les porte-manteaux, *or* les pistolets d'embarcation.

day's work (of a ship).	la route du vaisseau dans les 24 heures, *or* le point.
dead calm.	un calme plat.
— *eye.*	*cap* (*m.*) *de mouton.*
— *block.*	*poulie* (*f.*) *à moque.*
— *lights.*	*faux mantelets, m. pl., faux*
— *reckoning.*	*la route estimée.* [*sabords, m. pl.*
— *water.*	*le remous du sillage, m.*
— *works.*	*œuvres mortes, f. pl.* .
deck.	pont, *m.*
between —*s.*	*entre-pont, m.*
flush —.	*pont entier, m.*
gun —.	*batterie, f.*
main —.	*premier pont, m.*
lower —.	*premier pont, m.*
middle —.	*second pont, m.*
upper —.	*troisième pont, m.*
orlop deck.	. *faux pont, m.*
poop —.	*dunette, f.*
quarter —.	*le gaillard d'arrière.*
spar —.	*pont sur montants.*
to clear the —*s.*	*faire branle-bas.*
to —.	*ponter.*
two —*er.*	*navire à deux ponts.*
three —*er.*	*navire à trois ponts.*
deep sea lead.	la grande sonde.
deflection (of needle).	déclinaison, *f.* ⍟
demurrage.	surestaries, *f. pl.*
departure.	le chemin est ou ouest.
depth.	profondeur, *f.*
— *of hold.*	*creux, m.,* or *creux sur payol.*
derelict.	épave, *f.*
derrick.	un mât de rechange.
dingy.	youyou, *m.*
dip (of needle).	l'inclinaison, *f.*

dipping-needle.	une aiguille d'inclinaison.
disembark (to).	débarquer.
disembarkment.	débarquement, *m.*
distress (in).	en détresse.
to hoist a flag of —.	*mettre son pavillon en berne.*
signal of —.	*signal de détresse.*
dive (to).	plonger.
diver.	plongeur, *m.*
diving-bell.	une cloche à plongeur.
dividers.	compas (*m.*) à pointes sèches.
dock.	bassin, *m.*
dry —.	*bassin à sec.*
floating —.	*bassin à flot.*
— *dues.*	*les droits de bassin.*
to —.	*faire passer au bassin.*
dockyard.	un arsenal de la marine, chan-
dog vane.	penon, *m.* [tier, *m.*
dog watch.	petits quarts, *m. pl.*
dolphin striker.	un arc-boutant de martingale.
donkey-engine.	petit-cheval, *m.*
double (a cape) (to).	doubler.
downhaul.	hâle-bas, *m.*
draft (of water).	tirant d'eau, *m.*
drag.	drague, *f.*
draught.	*see* draft.
dress (to) ship.	pavoiser.
drift (to).	dériver.
drift.	dérive, *f.*
drive (to).	dériver.
driven ashore.	jeté à la côte.
driver (sail).	flèche-en-cul, *f.*
drizzle (to).	brouillasser.
drizzling rain.	pluie fine, *f.*
drop down with the tide (to).	descendre en jusant.
duck (canvass).	toile (*f.*) à voiles.

dunnage.	fardage, *m.*
dusk.	la brune.
duty.	le service.
to be on —.	*être de service.*
to be off —.	*ne pas être de service.*
the officer on —.	*l'officier de service.*
duties (custom).	droits, *m. pl.*
ease (a cable) (to).	filer (un cable).
ease off!	larguer !
— *her !*	*doucement !*
east.	est ; pronounce the *t.*
the —.	*l'orient, m., l'est, m.*
— *wind.*	*le vent d'est.*
— *ward.*	*à l'est.*
ebb.	reflux, *m.*
— *tide.*	*jusant, m.*
eccentric (of engine).	l'excentrique, *f.*
eddy.	remous, *m.*
edge away (to).	s'éloigner.
embargo.	embargo, *m.*
embark (to).	embarquer.
— (*of one's-self*).	*s'embarquer.*
embarkation.	embarquement, *m.*
engine.	machine, *f.*
— *room.*	*la chambre des machines.*
fire —.	*une pompe à incendie.*
high-pressure —.	*une machine à haute pression.*
low-pressure —.	*machine à basse pression.*
triple-expansion —.	*machine à triple expansion.*
quadruple-expansion —.	*machine à quadruple expansion.*
100 horse —.	*machine de cent chevaux.*
engineer.	mécanicien, *m.*
ensign.	le pavillon de poupe.
evolutions.	évolutions, *f.*, manœuvres, *f. pl.*
fleet —.	*manœuvres d'escadre.*
eye.	œil, *m.*

flemish eye.	*œil à la flamande.*
eyelet hole.	un œil de pie.
fair (weather).	beau temps.
— *wind.*	*bon vent.*
fathom.	brasse,* *f.*
to feather the oar.	mettre la rame à plat.
feathering paddle.	une aube articulée.
fender (of boat).	défense, *f.*
ferry-boat.	bac, *m.*
to — *over.*	*passer.*
—*man.*	*passeur, m.*
fid (of mast).	clé, *f.*
— (*splicing*).	*épissoir, m.*
to —.	*mettre en clé.*
fiddle block.	une poulie en violon.
fife rail.	râtelier, *m.*
fire.	feu, *m.*
to light the —*s.*	*allumer les feux.*
to bank the —.	*conserver les feux allumés.*
— *plug.*	*une prise d'eau.*
to — (*a gun*).	*tirer un coup de canon.*
fire !	feu !
fish davit.	le bossoir de traversière.
— *tackle.*	*traversière, f.*
to — (*the anchor*).	*traverser l'ancre.*
fisherman.	pêcheur, *m.*
fishing.	la pêche.
— *boat.*	*bateau pêcheur.*
flag.	pavillon, *m.*
— *ship.*	*vaisseau amiral, m.*
— *of truce.*	*le pavillon parlementaire, le drapeau blanc.*
— (*half-mast high*).	*le pavillon en berne,* or *à mi-mât.*
to strike the —.	*amener le pavillon.*
— *of truce* (*the bearer of a*).	*un parlementaire, m.*

---

* The French 'brasse' is equal to 1·624 metre, or very nearly 5 ft. 4 inches. An English 'fathom' is equal to 1·829 metre=6 feet..

flag-staff.	*le mât de pavillon.*
fleet.	flotte, *f.*
float (to).	surnager.
a —.	*flotteur, m.*
flood.	marée, *f.*
— *tide.*	*marée, f., flot, m.*
flotsam.	le jet de la mer.
flow (to). ·	couler.
fly-boat.	flibot, *m.*
foam.	écume, *f.*
foamy.	écumeu-x, -se.
F. O. B. (free on board).	sous vergues.
focus.	foyer, *m.*
to focus.	*mettre au point.*
focussing.	la mise au point.
fog.	brume, *f.* [*à bouquin.*
— *horn.*	*une trompe de brume, un cornet*
— *signal.*	*un signal de brume.*
it is foggy.	il y a de la brume.
foot.	pied, *m.* (*see* Table III.).
— *rope.*	*marchepied, m.*
force-pump.	une pompe foulante.
fore.	la misaine.
— *braces.*	*bras de misaine.*
— *cabin.*	*la chambre d'avant.*
— *course.*	*misaine, f.*
— *chains.*	*porte haubans de misaine.*
— *and main top sails.*	*huniers, m. pl.*
— *mast.*	*mât (m.) de misaine.*
— *royal.*	*petit cacatois.*
— *royal mast.*	*petit mât de cacatois.*
— *sail.*	*misaine, f.*
— *stay.*	*un étai de misaine.*
— *sky-sail.*	*petit cacatois volant.*
— *spencer.*	*misaine goëlette.*

fore stay-sail.	*la trinquette.*
— *tackle.*	*le gréement de l'avant.*
— *top.*	*hune de misaine.*
— *top-sail.*	*petit hunier.*
— *top-mast.*	*petit mât de hune.*
— *top-mast stay-sail.*	*petit foc.*
— *top-gallant mast.*	*mât de petit perroquet.*
— *top-gallant sail.*	*petit perroquet.*
— *yard.*	*vergue de misaine.*
— *and aft.*	*de l'avant à l'arrière.*
forecastle.	le poste de l'équipage.
forecastle.	le gaillard d'avant.
foreland.	promontoire, *m.*
fore-tackle.	la candelette.
fork and lashing eyes.	fourche de collier d'étai.
forward.	à l'avant.
foul.	engagé.
to — (*of the anchor*).	*surjaler, engager une ancre.*
to — (*hawse*).	*avoir des tours de chaîne.*
frame (of ship).	membrure, *f.*
freight.	cargaison, *f.*
— (*price*).	*frêt, m.*
to —.	*fréter.*
freighter.	affréteur, *m.*
frigate.	frégate, *f.*
frost.	la gelée.
fuel.	combustible, *m.*
funnel (of steamer).	cheminée, *f.*, tuyau, *m.*
furl (to) a sail.	serrer une voile.
furnace.	fourneau, *m.*
futtock shrouds.	gambes (*f.*) de revers.
gaff.	pic, *m.*
at the end of the —.	*au pic.*
spanker —.	*la corne.*

gaff sail.	*voile à corne.*
— halliards.	*drisses du pic.*
— top-sail.	*flèche d'artimon, f.*
gale.	tempête, *f.*
galleys.	cuisine, *f.*
gammonings of the bowsprit.	liures (*f.*) de beaupré.
gangway.	passavant, *m.*
— port.	*la portière.*
gantline.	cartahu, *m.*
garboard strake.	gabord, *m.*
garnet.	bredindin, *m.*
clew —.	*un cargue-point.* [ferlage.
gasket.	jarretière, *f.* ; un raban de
gauge.	jauge, *m.*
— (*draft of water*).	*le tirant d'eau.*
steam —.	*manomètre, m.*
gear.	drisse, *f.* ; gréement, *m.*
to throw out of —.	*arrêter.*
gig.	yole, *f.* ; youyou, *m.*
gin-block.	chape, *f.*
girt-line.	cartalin, *m.*
glass (spy).	télescope, *m.*
weather —.	*baromètre, m.*
hour —.	*sablier, m.*
go about (to).	virer de bord vent devant.
— *ahead!*	*en avant!*
— *astern!*	*en arrière!*
going free.	*courant largue.*
let her go off!	*laissez arriver!*
to go into dock.	*se radouber.*
to — (*down a river*).	*descendre, aller en aval.*
to — (*up a river*).	*monter, aller en amont.*
goods.	marchandises, *f. pl.*
goose neck.	potence, *f.*
grapnel.	grappin, *m.*

grappling-chains.	grappins, *m. pl.*
gratings.	caillebotis, *m.*
graving dock.	bassin (*m.*) de radoub.
gripe.	brion, *m.*
grommet.	estrope (*f.*) à ersiau.
guard ship.	vaisseau amiral.
guesswarp.	faux bras élongé en créance.
gun (cannon).	canon, *m.*
— *boat.*	*canonnière, f.*
— (*rifle*).	*fusil, m.*
rifled —.	*canon rayé.*
— *barrel.*	*canon de fusil.*
— *carriage.*	*affût, m.*
— *cotton.*	*fulmi-coton, m.*
— *deck.*	*batterie, f.*
— *port.*	*sabord, m.*
— *room.*	*fausse sainte barbe, f.*
— —.	*le poste des aspirants.*
— — *officers.*	*sous officiers.*
— *shot (distance).*	*à portée de fusil, de canon.*
— *tackle purchase.*	*itague, f.*
morning —.	*le coup de canon de diane.*
evening —.	*le coup de canon de retraite.*
gunpowder.	la poudre à canon.
gunsmith.	armurier, *m.*
gunwale.	plat-bord, *m.*
guy.	hauban de bout dehors.
— *rope.*	*maroquin, m.*
— *tackle.*	*palan (m.) de suspente.*
gybe (to).	gambéyer.
hail.	la grêle.
— (*within*).	*à portée de voix.*
halliard.	drisse, *f.*
hammer.	marteau, *m.*

handsomely !	en douceur !
handspike.	anspect, *m.*
hands (the) on board.	les hommes.
all — *ahoy !*	*tout le monde sur le pont !*
hank.	rocambeau, *m.*
harbour.	port, *m.*
— *dues.*	*droits de mouillage.*
— *master.*	*capitaine de port.*
— *of refuge.*	*port de refuge.*
tidal —.	*port de marée.*
hatches.	écoutilles, *f. pl.*
hatchway.	écoutille, *f.*
haul (to).	hâler.
— *aft !*	*bordez !*
— *down !* (*sails*).	*amenez !*
— *in !*	*hâlez à bord !*
— *out !* — *up !*	*hissez !*
— *taut !*	*raidissez ! embraquez !*
hauling to the wind.	serrer le vent.
haulyard.	drisse, *f.*
haven.	havre, *m.,* ; port, *m.*
hawse hole.	écubier, *m.*
— *plug.*	*tampon, m.*
hawser.	haussière, *f.*
— (*small*).	*faux bras, m. pl.*
towing —.	*remorque, f.*
— *bend.*	*nœud de plein poing, m.*
haze.	brume, *f.*
head (of a ship).	l'avant, *m.* ; le cap.
mast —.	see under *mast.*
figure —.	*guibre, m. ; éperon, m.*
— *wind.*	*vent debout.*
headland.	pointe, *f.*
health (bill of).	patente de santé, *f.*
— (*board of*)	*le bureau de la santé.*

heart.
moque, *f.*

heave (to).
jeter, lever.

— *away.*
virer.

— *down.*
virer en carène.

— *in at the capstan.*
virer au cabestan.

— *to.*
mettre en panne.

heave up the anchor!
dérapez!

heavily laden.
pesamment chargé.

heel (of a mast).
pied, *m.*

— *rope.*
braguet, m.

helm.
gouvernail, *m.*

the ship answers her —.
le navire gouverne bien.

the man at the —.
l'homme de barre.

— *port.*
le trou de jaumière.

mind the —!
attention à la barre!

put the — up!
arrive un peu!

helmsman.
timonnier, *m.*

high water.
marée haute, *f.*

— — *mark.*
le niveau des hautes eaux.

hitch.
nœud, *m.*

blackwall —.
nœud de bec d'oiseau.

double — —.
nœud de bec d'oiseau double.

bowline —.
nœud de chaise.

cat's-paw —.
gueule de raie.

clove —.
demi-clefs à capeler.

half —.
demi-clef, f.

— — *and timber hitch.*
demi-clef et barbouquet.

— — *and seizing.*
ajut avec demi-clef et amarrage.

marling —.
transfilage avec demi-clef.

marling-spike —.
nœud de griffe, patte (f.) de chat,
nœud de trésillon.

rolling —.
laguis, m., nœud de bosse, m.

timber —.
nœud de bois.

hitch (to).
nouer, amarrer.

H.M.S.
Vaisseau de sa Majesté.

hogged.	arqué.
hogging.	passer le goret.
hogshead.	barrique, *f.*
hoist (of a sail).	guindant, *m.* (de voile).
to —.	*hisser.*
— (*with a winch*).	*guinder.*
— *in* (*to*).	*embarquer.*
— *out* (*to*).	*débarquer.*
— *taut up !*	*étarquez !*
— *away !*	*hissez !*
hoisting.	guindage, *m.*
hold.	cale, *f.*
depth of —.	*le creux sur payol.*
hold on !	tenez bon !
homeward bound.	sur son retour.
hook.	croc, *m.*
— (*boat*).	*gaffe, f.*
fish —.	*hameçon, m.*
horse-power.*	*cheval vapeur, m.*
hose (fire).	une manche à eau.
houseline.	merlin, *m.*
housing a mast.	caler un mât.
hove to.	en panne.
hulk.	ponton, *m.*
hull.	coque, *f.* ; carène, *f.*
hurricane.	ouragan, *m.*
husband (ship's).	gérant, *m.*
ice.	glace, *f.*

---

* The English horse-power is equivalent to the work done by continuous exertion, at the rate of 33,000 lbs. raised through one foot in one
minute. *Cheval vapeur* is now equivalent to 100 kilogrammètres. The
kilogrammètre is equal to the work done by the continuous exertion
necessary to raise one kilogramme, one mètre, in one second. The
English horse-power is equal to seventy-six kilogrammètres.

ice-field.	*un banc de glace.*
— *berg.*	*banquise, f.*
import duty.	le droit d'entrée.
— *commerce.*	*commerce (m.) d'importation.*
inch.	pouce (*see* Table III.).
Indiaman (East).	navire des Indes.
inlet.	un bras de mer.
insurance.	assurance, *f.*
Lloyd's —.	*assurance maritime, f.*
invoice.	facture, *f.*
iron.	fer, *m.*
galvanized —.	*fer galvanisé, m.*
hoop —.	*du feuillard.*
cast —.	*de la fonte.*
pig —.	*gueuse, f.*
sheet —.	*de la tôle, f.*
wrought —.	*du fer forgé.*
— *clad.*	*cuirassé, blindé.*
— *wire.*	*du fil de fer, m.*
island.	île, *f.*
isthmus.	isthme, *m.*
jacket (of engine).	chemise, *f.*
jib.	foc, *m.*
flying —.	*clin foc, m.*
flying — *guys.*	*martingale, f.*
standing —.	*grand foc, m.*
— *boom.*	*bout dehors de foc.*
flying — *boom.*	*bout dehors de clin foc.*
— *stay.*	*draille, f.*
jigger.	tapecul, *m.*
jury-mast.	mât de fortune.
kedge.	ancre à jet.
kedging.	se touer.

keel.	quille, *f.*
false —.	*fausse quille.*
keelson.	carlingue, *f.*
keep her away !	laissez arriver ! laissez porter.
kentledge.	gueuse, *f.*
knees.	courbes, *f. pl.*
knight heads.	les apôtres.
knot.	nœud,* *m.*
to run 12 knots an hour.	*filer 12 nœuds à l'heure.*
how many knots do we run?	*combien filons-nous ?*
to tie a —.	*faire un nœud.*
running bowline —.	*nœud d'anguille.*
slip —.	*nœud coulant.*
sheet —.	*nœud d'écoute.*
fisherman's bend —.	*nœud de grappin.*
shroud —.	*nœud de hauban.*
buoy-rope —.	*nœud d'orin.*
reef —.	*nœud plat.*
figure of 8 —.	*nœud d'arrêt.*
half-crown —.	*amarrage en étrive.*
horse-shoe —.	*nœud de griffe.*
Matthew Walker —.	*nœud de ride.*
standing Turk's head —.	*bonnet turc.*
stopper —.	*cul de porc double.*
selvagee strop —.	· *erse en bitord.*
wall —.	*cul de porc simple.*
two bowlines —.	*nœud d'agui.*
cat's-paw —.	*une gueule de raie.*

---

* The French nautical mile, or knot, like the English nautical mile, or knot, is the length of a minute of latitude. Though it is different for every latitude, it is considered to be equal to 1852 metres, that is, 6076·40 feet. As the Admiralty knot is equal to 6080 feet, the difference between the French and the English nautical miles is then equal to 3·60 feet, a difference which it is hardly necessary to take into consideration except for very long distances.

ladder.	échelle, *f.*
lading (bill of).	connaissement, *m.*
lamp-oil.	huile (*f.*) à brûler.
land.	terre, *f.* [*terre* (*of goods*).
to —.	*débarquer* (*passengers*), *mettre à*
landing.	débarquement, *m.*
— *place or stage.*	*débarcadère, m.*
laniard or lanyard.	aiguillette, *f.*, ride, *f.*
lantern.	lanterne, *f.*
— (*of lighthouse*).	*fanal, m.*
larboard (*see* port).	babord, *m.*
latch.	linguet, *m.*
latitude.	latitude, *f.*
to launch.	mettre à l'eau, lancer.
a —.	*chaloupe, f.*
lay days.	jours de planche.
to lay up.	désarmer, désemparer.
lead.	plomb, *m.*
the — (*for sounding*).	*sonde, f.*
red —.	*minium, m.*
white —.	*du blanc de céruse.*
to heave the —.	*jeter la sonde.*
deep-sea —.	*grande sonde.*
leading seaman.	gabier, *m.*
leadsman.	le sondeur.
to leak.	faire eau.
to spring a —.	*faire une voie d'eau.*
leakage.	voie d'eau, *f.*
lee (on the).	sous le vent.
— *side.*	*côté sous le vent.*
brought by the —.	*empanné.*
— *shore.*	*terre sous le vent.*
— *tide.*	*marée portant sous le vent.*
— *way.*	*dérive, f.*
leech (of a sail).	chute, *f.*

leech line.	*cargue bouline, f.*
leeward (to).	sous le vent.
keep to — of —.	*arrivez pour —.*
lens.	lentille, *f.*
let go !	larguez !
letter of mark.	lettre de marque.
lieutenant (first).	lieutenant de vaisseau.
sub —.	*un enseigne.*
life belt.	ceinture (*f.*) de sauvetage.
— boat.	*un canot de sauvetage.*
— buoy.	*bouée (f.) de sauvetage.*
lifts.	balancines, *f. pl.*
light house.	phare, *m.*, feu, *m.*
— ship.	*phare flottant, m.*
floating —.	*un feu flottant.*
— house keeper.	*gardien de phare.*
first order —.	*un feu de premier ordre.*
fixed —.	*un feu fixe..*
revolving —.	*un feu tournant.*
*occulting —.**	*feu à éclipses.*
flashing —.	*feu à éclats.*
à flash every minute.	*un éclat de minute en minute.*
electric —.	*lumière électrique.*
to burn a — (signal).	*brûler une amorce.*
twinkling —.	*feu scintillant.*
flare-up —.	*feu à éclats.*
mast head —.	*feu de tête de mât.*
port bow —.	*feu de babord.*
starboard bow —.	*feu de tribord.*
signal —s.	*feux de signaux.*
a red —.	*un feu rouge.*
a green —.	*un feu vert.*

· * Always known by that name in the merchant service. The same as *intermittent light.*

a *white light.*	*un feu blanc.*
a — *on the port bow!*	*un feu par le bossoir de babord!*
a — *ahead!*	*un feu devant!*
lighter.	allège, *f.*
limbers.	anguilliers, *m. pl.*
— (*boards*).	*paracloses, pl.*
line.	corde, *f.*, amarre, *f.*
loading.	chargement, *m.*
loadstone.	aimant, *m.*
log (to measure speed).	loch, *m.*
to heave the —.	*jeter le loch.*
— *book.*	*journal de bord, m.*
longitude.	longitude, *f.*
long stay.	virer à long pic.
loof.	lof, *m.*
look-out (the).	vigie, *f.*
to be on the —.	*être en vigie.*
to keep a good —.	*veiller - bien, faire bon quart, avoir l'œil au bossoir.*
look out!	veille!
low water.	la marée basse.
— — *mark.*	*le niveau des basses eaux.*
lower mast.	bas-mât, *m.*
— *yard.*	*basse vergue, f.*
to — *a boat.*	*mettre une embarcation à la mer.*
to —.	*baisser.*
— *away.*	*amener.*
lubber's point.	ligne de foi.
to luff.	loffer.
luffer boards.	auvents articulés, *m. pl.*
lug-sail.	une voile à bourcet.
dipping —.	*voile à bourcet gambéyant.*
standing —.	*voile à bourcet fixe.*
lugger.	lougre, *m.* ; chasse-marée, *m.*
lull.	accalmie, *f.*

lump (in a dockyard).	une chaloupe de port.
lying to.	être à la cape.
magnet.	aimant, *m.*
mail steamer.	paquebot-poste, *m.* [*m. pl.*
main chains.	porte haubans de grand mât,
— *course.*	*grand'voile, f.*
— *mast.*	*grand mât, m.*
— *royal.*	*grand cacatois, m.*
— *sail.*	*grand'voile, f.*
— *sheet.*	*grand'écoute, f.*
— *stay-sail.*	*pouilleuse, f.*
— *top.*	*grand' hune, f.*
— *top-sail.*	*grand hunier, m.*
— *top-mast.*	*grand mât de hune, m.*
— *top-gallant mast.*	*mât de grand perroquet, m.*
— *top-gallant sail.*	*grand perroquet, m.*
— *yard.*	*grand'vergue, f.*
making a sternboard.	culer.
manger.	écubier de pont, *m.*
man-of-war.	bâtiment de guerre, *m.*
to man (the yards).	monter les vergues.
map.	carte, *f.*
— *of the world.*	*mappemonde, f.*
marine (a soldier).	soldat de marine.
royal — light infantry.	*l'infanterie de marine.*
royal — artillery.	*l'artillerie de marine.*
mariner's compass.	la boussole.
marline spike.	épissoir, *m.*
— *hitch.*	*nœud de trésillon, m.*
mast.	mât, *m.*
at head of the —.	*aux barres de cacatois, f. pl.*
— *head rigging.*	*capelage, m.*
— *head rope.*	*guinderesse, f.*
three —cd.	*à trois mâts.*

a *three master.*	*un trois mâts,* m.
masting.	mâter.
master mariner.	capitaine au long cours.
mean time, G.M.T.	temps moyen, m., T.M.
Mercator's charts.	cartes réduites, f.
merchantman.	navire marchand, m.
merchant vessel.	batiment (m.) de commerce.
ante-meridiem, A.M.	avant-midi, M.
post — (*P.M.*).	*après-midi* (*soir*), S.
midshipman.	aspirant de marine, m.
midships.	par le travers.
milky way.	la voie lactée.
mine.	mine, f.
submarine —.	*mine sous-marine,* f.
miss (to) stays.	manquer à virer.
mizen.	misaine, f.
—- *mast.*	*mât d'artimon.*
— *chains.*	*porte hauban d'artimon.*
— *mast-cap.*	*chouquet* (m.) *d'artimon.*
— *royal.*	*cacatois de perruche.*
— *royal-mast.*	*mât de cacatois de perruche.*
— *top.*	*la hune d'artimon.*
— *lug.*	*tapecul,* m.
— *top-mast.*	*mât de perroquet de fougue.*
— *top-sail.*	*perroquet de fougue.*
— *tap-gallant mast.*	*mât de perruche.*
— *top-gallant sail.*	*perruche,* f.
monsoon.	mousson, f.
moon.	la lune.
age of the —.	*l'âge de la lune.*
new —.	*la nouvelle lune.*
first quarter.	*le premier quartier.*
full —.	*la pleine lune.*
last quarter.	*le dernier quartier.*
moored (to be).	être affourché.

mooring.	amarrage, *m.*
—s.	corps mort, *m.*
— buoy.	coffre, *m.*
— ring.	organeau, *m.*
mop.	vadrouille, *f.*
mouse a hook (to).	moucheter un croc.
muzzle (of a gun).	la bouche (d'un canon).
nautical almanac.	la connaissance des temps, les éphémérides maritimes.
naval school.	école de marine, *f.*
— cadet.	*un élève de l'école navale.*
navy.	marine, *f.*
royal —.	*la marine militaire.*
merchant —.	*la marine marchande.*
the English —.	*la marine anglaise.*
the French —.	*la marine française.*
neap-tide.	marée de morte eau, *f.*, marée
needle (compass).	la boussole. [de quadrature.
net.	filet, *m.*
netting (of a ship).	bastingage, *m.*
— needle.	*navette, f.*
— pin.	*moule, m.*
no higher.	pas plus près.
noon.	midi, *m.*
north.	nord, *m.*
nothing off!	n'arrivez pas!
oakum.	étoupe, *f.*
oar.	aviron, *m.*, rame, *f.*
bow —.	*aviron de l'avant.*
stroke —.	*aviron de l'arrière.*
—s!	*lève rames!*
to lie on the —s.	*rester sur les avirons.*
back the —s!	*sciez! nage a culer!*

bend to your oars!	*nagez de long!*
ship your —s!	*armez!*
unship your —s!	*désarmez!*
off!	au large!
lying —.	*à la hauteur de.*
to get —.	*gagner le large.*
officer.	officier, *m.*
naval —.	*officier de marine.*
offing (in the).	au large.
orlop deck.	faux pont, *m.*
outhaul.	drisse, *f.*
—rigger.	*boute lof, m.*
overhaul (to).	visiter.
— (a sail).	*affaler.*

packet boat.	paquebot, *m.*
steam —.	*paquebot à vapeur, m.*
paddle (of canoe).	pagaie, *f.*
— of steamer.	*aube, f.*
— box.	*tambour, m.*
— steamer.	*un vapeur à aubes.*
— wheel.	*roue (f.) à aubes.*
feathering — —.	*roue à aubes articulées.*
painter.	bosse, *f.*
parallel motion.	le parallélogramme de Watt.
parbuckle.	trévire, *f.*
parcel a rope (to).	limander.
parrel.	racage, *m.*
partners.	les étambrais, *m.*
pass a stopper (to).	fouetter une bosse.
passage.	traversée, *f.*
— money.	*le prix de la traversée.*
passenger.	passag-er, -ère.
paul.	linguet, *m.*
to —.	*mettre les linguets à.*

pay the ship's head off (to).	faire retomber l'avant du navire.
paymaster.	officier payeur.
pay.	la solde.
full —.	*la solde entière.*
half —.	*la demi-solde.*
peak.	pic, *m.*
to —.	*apiquer.*
— *halliards.*	*drisses (f. pl.) du pic.*
pendant (flag).	flamme, *f.*
broad —.	*guidon, m.*
pendulum.	pendule, *m.*, balancier, *m.*
pennant.	pantoire, *f.*
periodic time.	révolution sidérale, *f.*
pier.	jetée, *f.*
— *dues.*	*le droit de jetée.*
pilot.	pilote, *m.*
— *boat.*	*bateau-pilote.*
pilotage fee.	frais de pilotage.
pinnace.	le grand canot.
pintle.	aiguillot, *m.*
pitch.	poix, *f.*, brai, *m.*
pitching (of a ship).	tangage, *m.*
point (of the compass).	une aire de vent.
— *a rope (to).*	*faire une queue de rat.*
pole.	pole, *m.*
— *star.*	*étoile polaire, f.*
poop.	dunette, *f.*
port (harbour).	port, *m.*
— *hole.*	*sabord, m.*
— *(side of ship).**	*babord, m.*

* In the French navy the words *babord* and *tribord*, 'port' and 'starboard,' have reference to the head of the ship, and not to the helm. Thus the command *babord!* 'port!' does not mean that the helm is to be put over to port, but that the head of the ship is to go to port. When the order refers specially to the tiller, the French officer com-

port!	*babord!*
on the — bow.	*par babord devant.*
on the — tack.	*babord amures.*
hard a —!	*babord tout!*
haul up the —s (to).	*ouvrir les sabords.*
lower the —s (to).	*fermer les sabords.*
powder.	poudre, *f.*
— *magazine.*	*la soute aux poudres.*
pressure gauge.	manomètre, *m.*
high —.	*à haute pression.*
low —.	*à basse pression.*
privateer.	un corsaire.
prow.	proue, *f.*
to pull (with an oar).	nager.
— *port!*	*nage babord!*
— *starboard!*	*nage tribord!*
to — stroke.	*donner la nage.*
pump.	pompe, *f.*
main —s.	*pompes royales, f.*
— *well.*	*l'archipompe.*
to work a —.	*manœuvrer une pompe.*
pumping engine.	une machine d'épuisement.
to put about (a ship).	virer.
— *back.*	*relâcher.*
to put into port.	*relâcher.*
— *in for.*	*faire route pour.*
quadrant.	un quart de cercle.
quarantine.	la quarantaine.
to perform —.	*faire la quarantaine.*
quarter-master.	gabier, quartier-maître.

---

mands, *babord la barre! tribord la barre!* In English, the order always
refers to the tiller supposed to be used, and the kind of steering-gear
employed makes no difference.

quarter-deck.	*le gaillard d'arrière.*
quay.	quai (*pronounced* kay), *m.*
rabbet.	râblure, *f.*
race (tide).	raz (*m.*) de marée.
raft.	radeau, *m.*
rail (of a ship).	lisse, *f.*
rain.	pluie, *f.*
drizzling —.	*bruine, f.*
heavy —.	*grosse pluie.*
— *awning.*	*taud, m.*
— *gauge.*	*pluviomètre, m.*
rainy.	pluvieu-x, -se.
ram (of a ship).	éperon, *m.*
ramrod (of rifle).	baguette, *f.*
— (*of cannon*).	*refouloir, m.*
ratlines.	enflêchures, *f. pl.*
ready about!	pare à virer!
reckoning.	estime, *f.*
dead —.	*la route estimée.*
red.	rouge.
reef (rock).	récif, *m.*
reef.	ris, *m.*
to take a —.	*prendre un ris.*
to shake out a —.	*larguer un ris.*
close reefed.	*au bas ris.*
— *bands.*	*bandes de ris.*
— *knot.*	*nœud plat.*
— *tackle.*	*palanquin, m.*
to reeve.	passer.
to refit.	se radouber.
revenue cutter.	patache, *f.*
ribs.	membres, *m. pl.*
rifle.	fusil, *m.*, carabine, *f.*
rig.	gréement, *m.*

to rig.	gréer.
rigged (the sails).	voilé.
square — ship.	un carré.
barque —.	gréé en barque.
de Horsey —.	gréé à la de Horsey.
rigging.	gréement, m., capelage, m.
standing —.	manœuvres dormantes, f.
running —.	manœuvres courantes, f.
lower —.	capelage de bas mât.
topmast —.	capelage de hune.
— of masts (the).	le capelage des mâts.
ring (of anchor).	cigale, f.
— (of moorings).	organeau, m.
rhumb.	rhumb, m.
river.	rivière, f.
up the —.	en amont.
down the —.	en aval.
in the —.	en rivière.
R. N.	de la marine royale.
roadstead.	rade, f.
roband.	une garcette d'envergure.
to roll (like a ship).	avoir du roulis.
rolling tackle.	un palan de roulis.
rope.	cordage, m., manœuvre, f.,
— (1 to 2 inches).	quarantenier, m. [corde, f.
bolt —.	ralingue, f.
foot — (yards).	marche-pied, m.
leech —.	une ralingue de chute.
— band.	un raban de tétière.
— maker.	cordier, m.
— yarn.	fil de caret, m.
tow —.	amarre, f.
hawser laid —.	une aussière en trois.
shroud laid —.	une aussière en quatre.
cable laid —.	grelin, m.

rose lashing.	aiguilletage, *m.*
royal yard.	la vergue de cacatois.
to row.	ramer.
rowlock.	tolletière, *f.*
row-port.	un sabord d'aviron.
rudder.	gouvernail, *m.*
— *case.*	*le trou de jaumière.*
rule of the road (the).	la règle pour gouverner.
to run (between two places).	faire le service de — à. . . .
to — aground (purposely).	*échouer.*
to run ashore.	*s'échouer.*
saddle.	racage, *m.*
safety valve.	la soupape de sureté.
sail.	voile, *f.*
to —.	*faire voile, naviguer.*
— (on a particular course).	*— cingler.*
to bend a —.	*enverguer une voile.*
to unbend a —.	*déverguer une voile.*
to carry plenty of —.	*porter de la toile.*
to shorten —.	*diminuer de voiles.*
to make more —.	*forcer de toile.*
gybing a —.	*coiffer une voile.*
to loose —.	*larguer les voiles.*
to lower a —.	*amener une voile.*
to get under —.	*appareiller.*
to set —.	*mettre à la voile.*
to set a —.	*établir une voile.*
to take a —.	*rentrer une voile.*
to trim the —s.	*orienter les voiles.*
to take in a —.	*serrer une voile.*
to reef a —.	*prendre un ris.*
— cloth.	*de la toile à voile.*
— loft.	*la voilerie.*
— maker.	*voilier.*

studding sail.	*bonnette.*
square —s.	*voiles carrées.*
stay —.	*voile d'étai.*
try —.	*voiles goëlettes.*
fore and aft —s.	*voiles auriques.*
head (of a sail).	*tétière, f.*
leech —.	*chute, f.*
luff —.	*lof, m.*
foot —.	*fond, m.*
clews —.	*les points d'écoute.*
tack —.	*amure, f.*
sheets —.	*écoutes, f.*
sailing.	navigation, *f.*
plane —.	*navigation à l'aide des cartes.*
traverse —.	*navigation à la bouline.*
rate of —.	*la vitesse du navire.*
great circle —.	*navigation orthodromique,* or *sur l'arc de grand cercle.*
— *vessel.*	*navire à voile.*
sailings (list of).	*navires en partance.*
sailor.	marin, *m.*
saloon.	grand salon.
— *passenger.*	*voyageur de première.*
salute.	salut, *m.*
to fire a —.	*faire un salut.*
to return a —.	*rendre un salut.*
a — *of 21 guns.*	*un salut de 21 coups de canon.*
samson's post.	épontille, *f.*
screw (of steamer).	hélice, *f.* [devant le temps.
to scud.	avoir le vent sous vergue, fuir
— *under bare poles.*	*courir à sec.*
scupper.	dalot, *m.*
scuttle (of a ship).	hublot, *m.*
to —.	*saborder, couler bas.*
sea.	mer, *f.*

arm of the sea.	*un bras de mer.*
a —.	*un coup de mer.*
at —.	*sur mer.*
high —s.	*la pleine mer.*
inland —.	*mer intérieure.*
open —.	*la pleine mer.*
to put to —.	*mettre à la mer.*
to ship a —.	*embarquer un coup de mer.*
to go to —.	*prendre la mer.*
to stand out to —.	*se tenir au large.*
to come from the —.	*venir de la mer.*
— chart.	*carte marine, f.*
— coast.	*la côte, le littoral.*
— man.	*homme de mer.*
— port.	*port de mer, m.*
— shore.	*côte, f.*
the state of the —.	*l'état de la mer.*
calm —.	*mer belle.*
chopping —.	*mer clapoteuse.*
high —.	*grosse mer.*
heavy —.	*mer houleuse.*
long —.	*mer longue.*
squally —.	*mer assez grosse.*
short —.	*mer courte.*
smooth —.	*mer plate.*
very smooth —.	*mer très belle.*
very heavy —.	*mer furieuse.*
the — is getting up.	*la mer se fait.*
the — is falling.	*la mer tombe.*
seaman.	marin, matelot, *m.*
seamanship.	manœuvre, *f.*, matelotage, *m.*
seizing.	amarrage, *m.*
flat —.	*amarrage à plat.*
throat —.	*amarrage plat double.*
racking —.	*amarrage en portugaise.*

sennit.	tresse anglaise, *f.*
French —.	*tresse, f.*
sentry.	factionnaire, *m.*
serve a rope (to).	fourrer une manœuvre.
setting up rigging.	rider le gréement.
sextant.	sextant, *m.*
shackle.	manille, *f.*
sheave hole (of mast).	caisse, *f.*
sheepshank.	une jambe de chien.
sheet (rope).	écoute, *f.*
— *tackle.*	*un palan de retenue.*
shelf-piece.	la bauquière.
shell (shrapnel).	obus, *m.*
— (*bomb*).	*bombe, f.*
shift (of the wind).	une saute de vent.
ship.	navire, *m.*, bâtiment, *m.*
a full-rigged —.	*un carré.*
on board —.	*à bord.*
—*'s company.*	*l'équipage, m.*
—*'s steward.*	*le commis aux vivres.*
— *of the line.*	*vaisseau de ligne.*
sailing —.	*navire à voile.*
steam —.	*navire à vapeur.*
screw —.	*navire à hélice.*
starboard wing —.	*le vaisseau de droite.*
port wing —.	*le vaisseau de gauche.*
the rear —.	*le vaisseau de queue,* or *serre-file.*
training —.	*un vaisseau école.*
training — *for naval cadets.*	*l'école de marine, f.*
training — *for boys.*	*l'école des mousses, f.*
— *broker.*	*courtier maritime.*
—- *builder.*	*constructeur de navires.*
— *chandler.*	*fournisseur de navires.*
— *owner.*	*armateur.*
—*'s husband.*	*le gérant du bord.*

ship's papers.	les papiers du bord.
— yard.	chantier, m.
shipment.	chargement, m.
shipping.	la marine marchande.
— agent.	expéditeur, m.
— charges.	frais de mise à bord.
— intelligence.	nouvelles de mer.
— office.	une agence maritime.
shore.	la terre, le littoral.
to shove off (a boat).	pousser au large.
shove off!	au large !
shrapnel.	obus, m.
shrouds.	haubans, m. pl.
sight (in).	en vue.
within —.	à portée de la vue.
to lose — of land.	perdre la terre de vue.
to take —s.	prendre des observations.
signal.	signal, m., signaux, pl.
sink (of a ship).	la sentine.
skipper.	patron, m.
skylight (of a ship).	claire-voie, f.
slab-lines.	fausses-cargues, f.
slack off!	mollissez !
sling (to).	élinguer.
butt —.	élingue double, f.
sloop.	sloop, m.
— of war.	corvette, f.
slops (of a sailor).	les effets d'habillement.
smack (fishing).	bateau pêcheur.
smuggler.	contrebandier, m.
sound.	sonde, f.
sounding.	sondage, m.
to try the —s.	chercher la sonde.
south.	sud (see compass).
southerly.	du sud, du midi.

spanker. — brigantine, *f.*

— *boom.* — *gui* or *bôme, f.*

— *gaff.* — *corne, f.*

— *peak halliard.* — *martinet, m.*

spar. — espars, *m.*

spare —s. — *drome, f.*

speed. — vitesse, *f.*

half —. — *demi-vapeur.*

full —. — *à toute vapeur.*

spider. — un arc-boutant de brasséyage.

splice. — épissure, *f.*

to —. — *épisser.*

eye —. — *œil épissé.*

long —. — *épissure longue, f.*

short —. — *épissure carrée, f.*

splicing-fid. — épissoir, *m.*

sponsons. — les jardins des tambours, *m.*

— *rims.* — *élongis des jardins, m.* ·[*dière*).

sprit-sail. — voile à livarde (formerly *civa-*

sprocket wheel. — hérisson, *m.*

— — *of the capstan.* — *la couronne.*

spun-yard. — bitord, *m.*, fil de caret, *m.*

squadron (of ships). — escadre, *f.*

square the yards ! — brassez carré !

squall. — rafale, *f.*

squally. — venteux.

staff and ball. — balise, *f.*

stanchion. — étançon, *m.*

quarter —s. — *jambettes, f. pl.*

stand by ! — attention ! soit paré à . . . . !

— *across.* — *mettre en travers.*

— *on.* — *suivre la même route.*

— *out to sea.* — *se tenir au large.*

— *to eastward.* — *faire route à l'est.*

standard (flag). — pavillon, *m.*

standard (timber).	*courbe, f.*
starboard.*	tribord.
hard a —!	*tribord tout!*
on the — bow.	*par tribord devant.*
— a bit!	*tribord un peu!*
staunch.	étanche.
stay.	*étai, m.*
back —s.†	*galhaubans, m.*
to be smart in —.	*virer facilement.*
to be in —s.	*être vent devant.*
to miss —s.	*manquer à virer.*
short —s.	*virer à pic.*
the ship is hove in —.	*le navire vire bien.*
steady (hoisting)!	doucement!
— (steering)!	*comme-ça!*
steam.	vapeur, *f.*
by —.	*à la vapeur.*
to get up —.	*chauffer.*
is — up?	*y-a-t-il de la pression?*
— is up.	*il y a de la pression.*
to shut off —.	*couper la vapeur.*
— boat.	*bateau à vapeur.*
— engine.	*machine à vapeur, f.*
— pressure gauge.	*manomètre m.*
— navigation.	*navigation à la vapeur.*
— ship.	*navire à vapeur, m.*
to —.	*jeter de la vapeur.*
to — full speed.	*aller à toute vapeur.*
to — away.	*s'éloigner.*
steamer.	vapeur, *m.*
to steer.	gouverner.
— clear of.	*éviter.*

* See note on 'port,' p. 37.
† For the names of stays, see under the name of each special one.

to steer close to the wind.	gouverner au plus près.
steerage.	timonerie, f.
— (fore).	l'avant, m.
— passenger.	passag-er, -ère, de l'avant.
— — on large steamers.	passag-er, -ère, d'entrepont.
— way.	sillage, m.
steering.	action de gouverner.
steersman.	timonier, m.
stem.	étrave.
from — to stern.	de l'avant à l'arrière.
stemson.	le marsouin d'avant.
stern.	l'arrière, m.
— ports.	sabords de retraite.
— post.	étambot, m.
in the — sheets.	à l'arrière.
— way.	aller en culant.
stevedore.	arrimeur, m.
steward.	le commis aux vivres.
—'s room.	la cambuse.
mess —.	le maître d'hotel.
stirrups.	étriers, m.
stoke-hole.	la chambre de chauffe.
stoker.	chauffeur, m.
stop !	stop !
to — (the engines).	stopper.
— her !	stoppez !
store-room (on board ship).	soute, f.
storm.	orage, m., ouragan, m.
stormy.	orageu-x, -se.
stove in.	défoncé.
to stow.	arrimer.
— (the anchor).	saisir.
stowage.	l'arrimage, m.
strand (shore).	plage, f., grève, f.
— (of a rope).	toron, m.

stranding.	échouement, *m.*
stress of weather.	gros temps, *m.*
strop.	estrope, *f.* [*nœuvre.*
to put a — on a rope.	frapper une erse sur une ma-
to put a — on a mast.	frapper une erse sur un mât.
studding sail.	bonnette, *f.*
lower —.	bonnette basse.
top-gallant —.	bonnette de hune.
— boom.	bout dehors de bonnettes.
submarine.	sous-marin, *m.*, sous-marine, *f.*
sun.	soleil, *m.*
— beam.	un rayon de soleil.
— rise.	le lever du soleil.
— set.	le coucher du soleil.
at — rise.	au lever du soleil.
at — set.	au coucher du soleil.
supercargo.	subrécargue, *m.*
surf.	ressac, *m.*
surge.	lame, *f.*
— (to).	choquer.
surgeon.	chirurgien, *m.*
swab.	faubert, *m.*
— (for gun).	écouvillon, *m.*
swamped.	empli.
swell.	houle, *f.*
there is some —.	il y a de la houle.
swinging boom.	tangon, *m.*
to lie at the —.	séjourner au tangon.
tack.	amure, *f.*
— (to make a).	faire une bordée, *f.*
the gaining —.	le bon bord.
to make short —s.	courir à petits bords.
down main —!	amure la grand'voile!
to be on the port —.	avoir les amures à babord.

E

to be on the starboard tack.	avoir les amures à tribord.
lee —.	amure de r.vers.
to raise tacks and sheets.	lever les lofs.
to —.	virer.
to —. about.	louvoyer, courir une bordée.
tackle (the) of a ship.	les apparaux, m.
luff —.	palan à croc, m. [caliorne, f.
— (apparel and furniture).	les agrès, m. pl., palan, m.,
fore —.	cartahu double.
top —.	caliorne, f.
deck —.	marguerite, f.
reef —.	palanquin, m.
rolling —.	palan de roulis, m.
taffrail.	couronnement, m.
tar.	goudron, m.
to —.	goudronner.
target.	cible, f.
to fire at a—.	tirer à la cible.
tarpaulin.	toile goudronnée, f.
tatoo.	la retraite.
to beat a —.	battre la retraite.*
telescope.	une lunette d'approche, une
tempest.	tempête, f. [longue vue.
tempestuous.	orageux.
tender.	patache, f., allége, f.
thermometer.	thermomètre, m.
thick (weather).	brumeux.
thimble.	cosse, f.
throat halliards.	une drisse de mât.
thrumb (to).	larder.
thunder.	tonnerre, m.
— storm.	orage, m.
thwarts.	les bancs de rameurs.

---

* Not to be confounded with *battre en retraite,* 'to retreat.'

tidal.	de marée.
— *harbour.*	*port à marée.*
tide.	marée, *f.*
flood —.	*le flot.*
ebb —.	*la marée descendante.* [*eau.*
spring —.	*marée de syzygie, marée de vive*
neap —.	*la morte-eau.*
the turn of the —.	*le changement de marée.*
weather —.	*marée portant au vent.*
tiller.	drosse (*f.*) de gouvernail.
— *ropes.*	*mèches, f.*
toggle.	cabillot, *m.*
ton.	tonneau (English ton = 1016
	kilos.).
tonnage.	jauge, *f.*
register —.	*jauge officielle.*
net register —.	*jauge nette.*
top.*	hune, *f.*
top-mast cross-tree.	les barres de perroquet.
top-sail.	hunier, *m.*
— *yard.*	*vergue d'hune.*
upper —.	*hunier volant.*
lower —.	*hunier fixe.*
mizen —.	*perroquet de fougue.*
buntlines —.	*retraite de hune.*
— *schooner.*	*brick goëlette, m.*
tornado.	ouragan, *m.*, cyclone, *m.*
torpedo.	torpille, *f.*
— *boat.*	*torpilleur, m.*
to plant a ground —.	*mouiller une torpille de fond.*
locomotive —.	*torpille automotrice.*
Whitehead —.	*la torpille Whitehead.*
electric —.	*torpille électrique.*

---

* See under the names of the different masts and sails.

to fish up a torpedo.	*relever une torpille.*
to explode a —.	*faire partir une torpille.*
clock-work —.	*torpille à mouvement d'horlo-*
tow (stuff).	étoupe, *f.* . [*gerie.*
to — (*a boat or ship*).	*remorquer.*
to take in —.	*prendre à la remorque.*
— *rope.*	*le faux bras.*
towage.	la remorque.
towing.	remorque, *f.*
— *hawser.*	*grelin, m.*
transom.	la barre d'arcasse.
tricing line.	suspensoir, *m.*
treenail.	gournable, *f.*
to trim the sails.	bien orienter les voiles.
trough of the sea.	le creux de la lame.
truck.	la pomme du mât.
main —.	*la pomme du grand mât.*
trysail (storm).	goëlette de cape, *f.*
tug.	remorqueur, *m.*
turret ship.	navire à tourelle.
to unbend (a sail).	déverguer.
to — (*a cable*).	*détalinguer.*
uniform.	uniforme, *m.*
full dress —.	*la grande tenue.*
undress —.	*la petite tenue, la tenue de service.*
summer dress —.	*la tenue d'été.*
winter dress —.	*la tenue d'hiver.*
union jack.	le pavillon anglais.
unlading.	déchargement, *m.*
unloading.	
days for —.	*jours de starie.*
unmoor.	désaffourcher.
unmooring.	démarrage, *m.*
unrig.	dégréer.

unship.	désarmer, débarquer.
— oars !	lève rames !
valve.	soupape, *f.*
safety —.	*soupape de sûreté.*
vane.	girouette, *f.*
variation (of the compass).	la variation du compas.
veer (a cable) (to).	filer.
to — away cable.	*filer du cable.*
the wind has —ed.	*le vent a changé.*
vessel.	vaisseau, *m.*
victuals.	les vivres, *m.*
wages.	la solde.
wake (of a ship).	sillage, *m.*
wales.	préceintes, *f.*
chain —.	*porte haubans, m. pl.*
ward-room.	le carré des officiers.
— officers.	*les officiers du carré.*
warrant officer.	maître, *m.*
watch.	quart, *m.*
to be on the —.	*faire le quart.*
dog —.	*les petits quarts.*
water.	eau, *f.*
— line.	*la ligne de flottaison.*
— logged.	*plein d'eau.*
— spout.	trombe, *f.*
— tight.	étanche.
— — bulkheads.	*cloisons étanches.* [*faire de l'eau.*
we called at Malta to —.	*nous avons relaché à Malte pour*
the ship makes —.	*le navire fait eau.*
waterman.	batelier, *m.*
wave.	vague, *f.*
way.	erre, *f.*
to get under —.	*appareiller.*

under way.	*appareiller en marche.*
to wear.	virer lof pour lof.
weather.	temps, *m.*
— (*to*).	*doubler.*
bad —.	*mauvais temps.*
beautiful —.	*beau temps.*
clear —.	*temps clair.*
cloudy —.	*temps nuageux.*
cold —.	*temps froid.*
damp —.	*temps humide.*
dark —.	*temps sombre.*
dry —.	*temps sec.*
fair —.	*beau temps.*
foggy —.	*temps brumeux.*
foul —.	*mauvais temps.*
fresh —.	*temps frais.*
hot —.	*temps chaud.*
mild —.	*temps doux.*
misty —.	*temps embrumé.*
rainy —.	*temps pluvieux.*
rough —.	*temps dur.*
squally —.	*temps à grains.*
stormy —.	*temps orageux.*
sultry —.	*temps lourd.*
threatening —.	*temps menaçant.*
unsettled —.	*temps incertain.*
wet —.	*temps humide.*
a change of —.	*un changement de temps.*
forecasts of —.	*prédictions du temps.*
the — *is improving.*	*le temps s'embellit.*
— *side.*	*le côté du vent.*
hard a — *!*	*arrive tout !*
in stress of —.	*par un gros temps.*
weigh (to be under).	être en marche.
to — *the anchor.*	*lever l'ancre.*

west.	ouest, *m.* [1]
whaler.	un baleinier.
wharfinger.	un garde quai.
wheel (of helm).	la roue du gouvernail.
the man at the —.	*l'homme de barre.*
whelps.	les taquets de guindeau.
wherry.	bac, *m.*, bachot, *m.*
whip a rope (to).	faire une sourliure.
whistle.	sifflet, *m.*
winch.	treuil, *m.*
wind.	vent, *m.*
to take the — *out of a sail.*	*déventer une voile.*
head —.	*vent debout.*
fair —.	*bon vent.*
to haul the —.	*serrer le vent.*
high —.	*grand vent.*
going before the —.	*avoir vent arrière.*
the — *has veered.*	*le vent à tourné.*
— *abeam.*	*vent de travers.*
a shift of —.	*une saute de vent.*
on a —.	*au plus près.*
steady —.	*vent fait.*
unsteady —.	*brise inégale.*
strong —*s.*	*de fortes brises.*
the — *has lulled.*	*le vent a molli.*
the — *is backing.*	*le vent redescend.*
the — *is due east.*	*le vent est franc est.*
the — *freshens.*	*la brise fraichit.*
the — *is getting up.*	*le vent se fait.*
how is the —?	*d'où vient le vent?*
northerly —.	*vent de nord.*
easterly —.	*vent d'est.*
southerly —.	*vent de sud.*
westerly —.	*vent d'ouest.*
trade —*s.*	*les vents alizés.*

NAUTICAL TERMS.

windlass.	guindeau, *m.*
windward (to).	au vent.
wings.	cloisons latérales, *f.*
work (to) a ship.	manœuvrer un bâtiment.
world (to sail round the).	faire le tour du monde.
worm (to).	congréer.
wreck.	naufrage, *m.*
yacht.	yacht, *m.*
yard.*	vergue, *f.*
top-sail —.	*la vergue d'hune.*
top gallant —.	*la vergue de perroquet.*
cross jack —.	*vergue de fortune.*
— *tackle purchase.*	*un palan de bout de vergue.*
— *arm.*	*un bout de vergue.*
the slings of a —.	*la suspente d'une vergue.*
to square the —*s.*	*dresser les vergues.*
square —*s !*	*brassez carré !*
yarn.	fil de caret, *m.*
spun —.	*bitord, m.*
yoke (of rudder).	la barre.
zenith.	le zénith.
zero.	zéro, *m.*

---

* Also see under the special names of yards.

TABLES.

FRENCH METRIC SYSTEM OF WEIGHTS AND MEASURES.

The system is decimal throughout.

The basis of the system is the *mètre*, which is equal to 39·3708 inches. The *mètre* is divided into ten parts, called *décimètres*; each *décimètre* into ten parts, called *centimètres*; and each *centimètre* into ten parts, called *millimètres*. Therefore, 1 mètre = 10 décimètres = 100 centimètres = 1000 millimètres.

Square and cubic measures are derived from the *mètre*. The sub-divisions of the square and cubic measures are the same as for the linear *mètre*, but naturally

1 square mètre	=		100	square décimètres.
,,	,,	=	10,000	,, centimètres.
,,	,,	=	1,000,000	,, millimètres.
1 cubic	,,	=	1,000	cubic décimètres.
,,	,,	=	1,000,000	,, centimètres.
,,	,,	=	1,000,000,000	,, millimètres.

The *litre*, used for measures of capacity, is equal to the contents of a cube whose side is 1 décimètre long. Therefore, a cubic décimètre is the same as 1 litre (not quite an English quart).

The *gramme*, which is the unit of weight, is equal to the weight of one cubic centimètre of pure water at the temperature of 4 degrees centigrade, weighed in Paris.

To form multiples, the words *deca* (10), *hecto* (100), *kilo* (1000), *myria* (10,000), are prefixed to the names of the particular measure. Thus *kilogramme* = 1000 grammes; *myriamètre* = 10,000 mètres. For the sub-multiples, the words *déci* ($\frac{1}{10}$), *centi* ($\frac{1}{100}$), *milli* ($\frac{1}{1000}$), are used in the same manner as the multiples, e.g. *décilitre* = $\frac{1}{10}$ of a litre; *centigramme* = $\frac{1}{100}$ of a gramme.

TABLE I.

Measures of Length.

		Inches.		Feet.		Yards.
Millimètre	=	0·03937	...	0·003281	...	0·0010936
Centimètre	=	0·39371	...	0·032809	...	0·010936
Décimètre	=	3·93708	...	0·328090	...	0·1093633
Mètre	=	39·37079	...	3·280989	...	1·0936331
Décamètre	=	393·70790	...	32·809892	...	10·9363306
Hectomètre	=	3937·07900	...	328·089917	...	109·3630956
Kilomètre	=	39370·79000	...	3280·899167	...	1093·6330556
Myriamètre	=	393707·90000	...	32808·991667	...	10936·3305556

Cubic, or Measures of Capacity.

		Cubic inches.		Cubic feet.		Pints.		Gallons.		Bushels.
Millilitre, or cubic centimètre	=	0·06103	...	0·000035	...	0·00176	...	0·0002201	...	0·0000275
Centilitre, 10 cubic do.	=	0·61027	...	0·000353	...	0·01761	...	0·0022010	...	0·0002751
Décilitre, 100 cubic do.	=	6·10271	...	0·003532	...	0·17608	...	0·0220097	...	0·0027512
Litre, or cubic décimètre ...	=	61·02705	...	0·035317	...	1·76077	...	0·2200967	...	0·0275121
Décalitre	=	610·27052	...	0·353166	...	17·60773	...	2·2009668	...	0·2751208
Hectolitre	=	6102·70515	...	3·531658	...	176·07734	...	22·0096677	...	2·7512085
Kilolitre	=	61027·05152	...	35·316581	...	1760·77341	...	220·0966767	...	27·5120846
Myrialitre	=	610270·51519	...	353·165807	...	17607·73414	...	2200·9667675	...	275·1208459

MEASURES OF WEIGHT.

	Grains.	Troy oz.	Avoirdupois lb.	Cwt.=112 lb.	Tons=20 cwt.
Milligramme	= 0·01543 ...	0·000032 ...	0·0000022 ...	0·0000000 ...	0·0000000
Centigramme	= 0·15432 ...	0·000322 ...	0·0000220 ...	0·0000002 ...	0·0000000
Décigramme	= 1·54323 ...	0·003215 ...	0·0002205 ...	0·0000020 ...	0·0000001
GRAMME	= 15·43235 ...	0·032151 ...	0·0022046 ...	0·0000197 ...	0·0000010
Décagramme	= 154·32349 ...	0·321507 ...	0·0220462 ...	0·0001968 ...	0·0000098
Hectogramme	= 1543·23488 ...	3·215073 ...	0·2204621 ...	0·0019684 ...	0·0000984
Kilogramme	= 15432·34880 ...	32·150727 ...	2·2046213 ...	0·0196841 ...	0·0009842
Myriagramme	= 154323·48800 ...	321·507267 ...	22·0462126 ...	0·1968412 ...	0·0098421

SQUARE, OR MEASURES OF SURFACE.

	Sq. feet.	Sq. yards.	Sq. perches.	Sq. roods.	Sq. acres.
Centiare, or square mètre	= 10·764299 ...	1·196033 ...	0·0395383 ...	0·0009885 ...	0·0002471
ARE, or 100 square mètres	= 1076·429934 ...	119·603326 ...	3·9538290 ...	0·0988457 ...	0·0247114
Hectare, or 10,000 sq. mètres	= 107642·993419 ...	11960·332602 ...	395·3828959 ...	9·8845724 ...	2·4711431

A French Quintal = 100 kilos = 1 cwt. 3 qrs. 24·46 lbs. = 220·46 lbs.

A French Ton = 1000 kilos = 19 cwt. 2 qrs. 20·61 lbs. = 2204·61 lbs.

TABLE II.—For the Conversion of Metric Weights and Measures into English.

Mètres into	yards.	Kiloms. into	miles.	yards.	Litres into	gals.	quarts.	Hecto-litres into	qrts.	bushels.	Kilogs. into	cwts.	qrs.	lbs.	oz.	Hectares into	acres.	roods.	poles.
1	1·094	1	0	1094	1	0	0·880	1	0	2·751	1	0	0	2	3¼	1	2	1	35
2	2·187	2	1	427	2	0	1·761	2	0	5·502	2	0	0	4	6½	2	4	3	31
3	3·281	3	1	1521	3	0	2·641	3	1	0·254	3	0	0	6	9¾	3	7	1	26
4	4·374	4	2	855	4	0	3·521	4	1	3·005	4	0	0	8	13	4	9	3	22
5	5·468	5	3	188	5	0	0·402	5	1	5·756	5	0	0	11	0¼	5	12	1	17
6	6·562	6	3	1282	6	1	1·282	6	2	0·507	6	0	0	13	3½	6	14	3	12
7	7·655	7	4	615	7	1	2·163	7	2	3·258	7	0	0	15	7	7	17	1	8
8	8·749	8	4	1709	8	1	3·043	8	2	6·010	8	0	0	17	10¼	8	19	3	3
9	9·843	9	5	1043	9	1	3·923	9	3	0·761	9	0	0	19	13½	9	22	0	38
10	10·936	10	6	376	10	2	0·804	10	3	3·512	10	0	0	22	0¾	10	24	2	34
20	21·873	20	12	753	20	4	1·608	20	6	7·024	20	0	1	16	1½	20	49	1	28
30	32·809	30	18	1129	30	6	2·412	30	10	2·536	30	0	2	10	2½	30	74	0	21
40	43·745	40	24	1505	40	8	3·215	40	13	6·048	40	0	3	4	3	40	98	3	15
50	54·682	50	31	122	50	11	0·019	50	17	1·560	50	0	3	26	3¾	50	123	2	9
60	65·618	60	37	498	60	13	0·823	60	20	5·072	60	1	0	20	4½	60	148	1	3
70	76·554	70	43	874	70	15	1·627	70	24	0·585	70	1	1	14	5¼	70	172	3	37
80	87·491	80	49	1251	80	17	2·431	80	27	4·097	80	1	2	8	6	80	197	2	38
90	98·427	90	55	1627	90	19	3·235	90	30	7·609	90	1	3	2	6½	90	222	1	24
100	109·363	100	62	243	100	22	0·039	100	34	3·121	100	1	3	24	7	100	247	0	18
200	218·727	200	124	487	200	44	0·077	200	68	6·242	200	3	3	20	15	200	494	0	37
300	328·090	300	186	730	300	66	0·116	300	103	1·362	300	5	3	17	6	300	741	1	15
400	437·453	400	248	973	400	88	0·155	400	137	4·483	400	7	3	13	14	400	988	1	33
500	546·816	500	310	1217	500	110	0·193	500	171	7·604	500	9	3	10	5	500	1235	2	11

TABLE III.

COMPARISON OF ENGLISH WITH METRICAL WEIGHTS AND MEASURES.

MEASURES OF LENGTH.

Inch (pouce)	=	0·02549	mètres.
Foot (pied)	=	0·30479	„
Yard	=	0·91438	„
Fathom (brasse)	=	1·82877	„
Pole or Perch (5½ yards)	=	5·02911	„
Furlong (220 yards)	=	201·16437	„
Mile (1,760 yards)	=	1609·31	„

SQUARE MEASURES.

Yard	=	0·83610	square mètres.
Rod	=	25·29194	„ „
Rood (1,210 square yards)	=	10·11677	„ „
Acre (4,840 square yards)	=	0·40467	hectares (100 ares).

CUBIC MEASURES.

Pint (⅛ of gallon)	=	0·5679	litre.
Quart (¼ of gallon)	=	1·1359	„
Imperial Gallon	=	4·543468	„
Peck (2 gallons)	=	9·086916	„
Bushel (8 gallons)	=	36·34766	„
Sack (3 bushels)	=	1·09043	hectolitre (100 litres).
Quarter (8 bushels)	=	2·90781	„
Chaldron (12 sacks)	=	13·08516	„

WEIGHTS.

Troy.	Grain (1/24 dwt.)	=	6·479895	centigrammes (1/100 gr.).
	Pennyweight (1/20 oz.)	=	1·555175	gramme.
	Ounce (1/12 lb. troy)	=	31·103496	„
	Pound (5,760 grs.)	=	373·241948	„
Avoirdupois.	Dram (1/16 oz.)	=	1·771846	„
	Ounce	=	28·349540	„
	Pound (7,000 grs.)	=	453·592645	„
	Cwt. (112 lbs.)	=	50·802	kilogrammes.
	Ton (20 cwt.)	=	1016·048	„

TABLE IV.

ENGLISH COINS AND BANK-NOTES.		FRENCH COINS AND BANK-NOTES.	
Monnaies anglaises et billets de banque.		Monnaies françaises et billets de banque.	

Gold.	frcs. cts.	Gold.	£. s. d.
Sovereign, £1	25 · 00*	Pièce de 100 frcs. ...	4 0 0†
Half-sovereign, 10s.	12 · 50	„ 50 „ ...	2 0 0†
		„ 40 „ ...	1 12 0†
Silver.		„ 20 „ ...	0 16 0
Crown, 5s.	6 · 25	„ 10 „ ...	0 8 0
Half-crown, 2s. 6d....	3 · 125	„ 5 „ ...	0 4 0†
Florin, 2s.	2 · 50	**Silver.**	
Shilling, 1s.	1 · 25	Pièce de 5 frcs. ...	0 4 0
Sixpence, 6d.	0 · 60	„ 2 „ ...	0 1 7½
Fourpence, 4d.	0 · 40	„ 1 „ ...	0 0 9¾
Threepence, 3d.......	0 · 30	„ 0·50 „ ...	0 0 4¾
Copper.		**Copper.**	
Penny, 1d.	0 · 10	10 centimes	0 0 1
Half-penny, ½d.	0 · 05	5 „	0 0 0½
Farthing..............	0 · 025	2 „	0 0 ·2
		1 „	0 0 ·1
Notes.		**Notes.**	
£5	125 · 00	50 francs............	2 0 0
£10.....................	250 · 00	100 „	4 0 0
£50.....................	1250 · 00	200 „	8 0 0
£100	2500 · 00	500 „	20 0 0
£1000..................	25000 · 00	1000 „	40 0 0

N.B.—The above equivalent values are given as nearly as possible, and without any regard to the fluctuation of the exchange.

* When changing English money into French, it is well to bear in mind that the rate of exchange is nearly always in favour of English money. The sovereign frequently fetches 25·25 frcs., and even as much as 25·37 frcs.

† Scarcely ever seen.

TABLE V.

THE ENGLISH BAROMETRIC SCALE IN FRENCH MILLIMÈTRES.

ENGLISH BAROMETER.		FRENCH BAROMETER.	
31	inches.	787·39	millimètres.
30·9	,,	784·85	,,
30·8	,,	782·31	,,
30·7	,,	779·77	,,
30·6	,,	777·23	,,
30·5	,,	774·69	,,
30·4	,,	772·15	,,
30·3	,,	769·61	,,
30·2	,,	767·07	,,
30·1	,,	764·53	,,
30	,,	761·99	,,
29·9	,,	759·45	,,
29·8	,,	756·91	,,
29·7	,,	754·37	,,
29·6	,,	751·83	,,
29·5	,,	749·29	,,
29·4	,,	746·75	,,
29·3	,,	744·21	,,
29·2	,,	741·67	,,
29·1	,,	739·13	,,
29	,,	736·59	,,
28·9	,,	734·05	,,
28·8	,,	731·51	,,
28·7	,,	728·97	,,
28·6	,,	726·43	,,
28·5	,,	723·89	,,
28·4	,,	721·35	,,
28·3	,,	718·81	,,
28·2	,,	716·27	,,
28·1	,,	713·73	,,
28	,,	711·19	,,
27·9	,,	708·65	,,
27·8	,,	706·11	,,
27·7	,,	703·57	,,
27·6	,,	701·03	,,
27·5	,,	698·49	,,
27·4	,,	695·95	,,
27·3	,,	693·41	,,
27·2	,,	690·87	,,
27·1	,,	638·33	,,
27	,,	685·79	,,

TABLE VI.

THE THERMOMETERS.

The Centigrade thermometer is universally used by scientific men, and it is the scale most generally employed on the continent. The Germans call it the 'Celsius thermometer.' The Réaumur scale is principally used in Russia. In these two scales the freezing point is denoted by o, and the boiling point respectively by 100 and 80. Therefore, 100 degrees Centigrade = 80 degrees Réaumur = 180 degrees Fahrenheit.

As the indications of the Fahrenheit scale begin at 32 degrees, and not at zero, we must remember that in converting from degrees Fahrenheit to Centigrade or Réaumur, 32 degrees must first be subtracted; whilst when we convert degrees Centigrade or Réaumur into Fahrenheit, 32 degrees must be added after the multiplication and division have been made.

It is important to remember that, since 100 degrees Centigrade = 80 degrees Réaumur = 180 degrees Fahrenheit,

$$5 \text{ degrees Centigrade}$$
$$= 4 \quad ,, \quad \text{Réaumur}$$
$$= 9 \quad ,, \quad \text{Fahrenheit.}$$

Example I. Express 59° Fahrenheit into Centigrade.

$9 : 5 :: (59 - 32) :$ Answer, that is, $\frac{5 \times 27}{9} = 15°$ Cent.

Example II. Express 95° Fahrenheit into Réaumur.

$9 : 4 :: (95 - 32) :$ Answer, that is, $\frac{4 \times 63}{9} = 28°$ Réaumur.

Example III.   Express 28° Réaumur into Fahrenheit.

4 : 9 :: 28 : Answer, that is, $\frac{9 \times 28}{4} = 63$, to which add 32
= 95° Fahrenheit.

Example IV.   Express 25° Centigrade into Fahrenheit.

5 : 9 :: 25 : Answer, that is, $\frac{9 \times 25}{5} = 45$, to which add 32
= 77° Fahrenheit.

Example V.   Express 25° Centigrade into Réaumur.

5 : 4 :: 25 : Answer, that is, $\frac{4 \times 25}{5} = 20$° Réaumur.

Example VI.   Express 28° Réaumur into Centigrade.

4 : 5 :: 28 : Answer, that is, $\frac{5 \times 28}{4} = 35$° Centigrade.

N.B.—If the number of degrees Fahrenheit to be converted into degrees of the other scales happened to be below 32, the results obtained would be negative, that is, below 0.

Example VII.   Express 23° Fahrenheit into Centigrade.

9 : 5 :: (23 − 32) : Answer, that is, $\frac{5 \times (-9)}{9} = -5$° Cent.

N.B.—See next page for a Table of the most common degrees of Fahrenheit into degrees of the Centigrade scale.

E

Here follows a table giving the degrees Centigrade corresponding to the degrees Fahrenheit most commonly used.

Fahrenh.	Centigrade.	Fahrenh.	Centigrade.	Fahrenh.	Centigrade.
— 10°	— 23·33	34°	1°11	78°	25·56
— 9	— 22·78	35	1·67	79	26·11
— 8	— 22·22	36	2·22	80	26·67
— 7	— 21·67	37	2·78	81	27·22
— 6	— 21·11	38	3·33	82	27·78
— 5	— 20·56	39	3·89	83	28·33
— 4	— 20·00	40	4·44	84	28·89
— 3	— 19·44	41	5·00	85	29·44
— 2	— 18·89	42	5·56	86	30·00
— 1	— 18·33	43	6·11	87	30·56
0	— 17·78	44	6·67	88	31·11
1	— 17·22	45	7·22	89	31·67
2	— 16·67	46	7·78	90	32·22
3	— 16·11	47	8·33	91	32·78
4	— 15·56	48	8·89	92	33·33
5	— 15·00	49	9·44	93	33·89
6	— 14·44	50	10·00	94	34·44
7	— 13·89	51	10·56	95	35·00
8	— 13·33	52	11·11	96	35·56
9	— 12·78	53	11·67	97	36·11
10	— 12·22	54	12·22	98	36·67
11	— 11·67	55	12·78	99	37·22
12	— 11·11	56	13·33	100	37·78
13	— 10·56	57	13·89	101	38·33
14	— 10·00	58	14·44	102	38·89
15	— 9·44	59	15·00	103	39·44
16	— 8·89	60	15·56	104	40·00
17	— 8·33	61	16·11	105	40·56
18	— 7·78	62	16·67	106	41·11
19	— 7·22	63	17·22	107	41·67
20	— 6·67	64	17·78	108	42·22
21	— 6·11	65	18·33	109	42·78
22	— 5·56	66	18·89	110	43·33
23	— 5·00	67	19·44	111	43·89
24	— 4·44	68	20·00	112	44·44
25	— 3·89	69	20·56	113	45·00
26	— 3·33	70	21·11	114	45·56
27	— 2·78	71	21·67	115	46·11
28	— 2·22	72	22·22	116	46·67
29	— 1·67	73	22·78	117	47·22
30	— 1·11	74	23·33	118	47·78
31	— 0·56	75	23·89	119	48·33
32	— 0·00	76	24·44	120	48·89
33	0·56	77	25·00	121	49·44

14, *Henrietta Street, Covent Garden, London ;* and
20, *South Frederick Street, Edinburgh.*

WILLIAMS AND NORGATE'S

LIST OF

𝔉rench, 𝔊erman, 𝔍talian, 𝔏atin and 𝔊reek,

AND OTHER

SCHOOL BOOKS AND MAPS.

𝔉rench.

FOR PUBLIC SCHOOLS WHERE LATIN IS TAUGHT.

Eugène (G.) The Student's Comparative Grammar of the French Language, with an Historical Sketch of the Formation of French. For the use of Public Schools. With Exercises. By G. Eugène-Fasnacht, French Master, Westminster School. 11th Edition, thoroughly revised. Square crown 8vo, cloth. 5s.

Or Grammar, 3s. ; Exercises, 2s. 6d.

"The appearance of a Grammar like this is in itself a sign that great advance is being made in the teaching of modern languages. . . . . The rules and observations are all scientifically classified and explained."—*Educational Times.*

"In itself this is in many ways the most satisfactory Grammar for beginners that we have as yet seen."—*Athenæum.*

Eugène's French Method. Elementary French Lessons. Easy Rules and Exercises preparatory to the "Student's Comparative French Grammar." By the same Author. 9th Edition. Crown 8vo, cloth. 1s. 6d.

"Certainly deserves to rank among the best of our Elementary French Exercise-books."—*Educational Times.*

Delbos. Student's Graduated French Reader, for the use of Public Schools. I. First Year. Anecdotes, Tales, Historical Pieces. Edited, with Notes and a complete Vocabulary, by Leon Delbos, M.A., of King's College, London. 3rd Edition. Crown 8vo, cloth. ' 2s.

—— The same. II. Historical Pieces and Tales. 3rd Edition. Crown 8vo, cloth. 2s.

Little Eugène's French Reader. For Beginners. Anecdotes and Tales. Edited, with Notes and a complete Vocabulary, by Leon Delbos, M.A., of King's College. 2nd Edition. Crown 8vo, cloth. 1s. 6d.

4000/9/88

Krueger (H.) Short French Grammar. 6th Edition. 180 pp.
12mo, cloth.   2s.

Victor Hugo. Les Misérables, les principaux Épisodes. With
Life and Notes by J. Boïelle, Senior French Master,
Dulwich College. 2 vols. Crown 8vo, cloth. Each 3s. 6d.

———— Notre-Dame de Paris. Adapted for the use of Schools
and Colleges, by J. Boïelle, B.A., Senior French Master,
Dulwich College. 2 vols. Crown 8vo, cloth. Each 3s.

Boïelle. French Composition through Lord Macaulay's English.
I. Frederic the Great. Edited, with Notes, Hints, and
Introduction, by James Boïelle, B.A. (Univ. Gall.), Senior
French Master, Dulwich College, &c. &c. Crown 8vo,
cloth.   3s.

Foa (Mad. Eugen.) Contes Historiques. With Idiomatic Notes
by G. A. Neveu. 3rd Edition. Crown 8vo, cloth. 2s.

Larochejacquelein (Madame de) Scenes from the War in the
Vendée. Edited from her Mémoirs in French, with
Introduction and Notes, by C. Scudamore, M.A. Oxon,
Assistant Master, Forest School, Walthamstow. Crown
8vo, cloth.   2s.

French Classics for English Schools. Edited, with Introduction
and Notes, by Leon Delbos, M.A., of King's College.
Crown 8vo, cloth.

No. 1. Racine's Les Plaideurs.   1s. 6d.
No. 2. Corneille's Horace.   1s. 6d.
No. 3. Corneille's Cinna.   1s. 6d.
No. 4. Molière's Bourgeois Gentilhomme.   1s. 6d.
No. 5. Corneille's Le Cid.   1s. 6d.
No. 6. Molière's Précieuses Ridicules.   1s. 6d.
No. 7. Chateaubriand's Voyage en Amérique.   1s. 6d.
No. 8. De Maistre's Prisonniers du Caucase and Lepreux
    d'Aoste.   1s. 6d.
No. 9. Lafontaine's Fables Choisies.   1s. 6d.

Lemaistre (J.) French for Beginners. Lessons Systematic, Practical and Etymological. By J. Lemaistre. Crown 8vo,
cloth.   2s. 6d.

Roget (F. F.) Introduction to Old French. History, Grammar.
Chrestomathy, Glossary. 400 pp. Crown 8vo, cl.   6s,

Kitchin. Introduction to the Study of Provençal. By Darcy B. Kitchin, B.A. [Literature—Grammar—Texts—Glossary.] Crown 8vo, cloth. 4*s.* 6*d.*

Tarver. Colloquial French, for School and Private Use. By H. Tarver, B.-ès-L., late of Eton College. 328 pp., crown 8vo, cloth. 5*s.*

Ahn's French Vocabulary and Dialogues. 2nd Edition. Crown 8vo, cloth. 1*s.* 6*d.*

Delbos (L.) French Accidence and Minor Syntax. 2nd Edition. Crown 8vo, cloth. 1*s.* 6*d.*

—— Student's French Composition, for the use of Public Schools, on an entirely new Plan. 250 pp. Crown 8vo, cloth. 3*s.* 6*d.*

Vinet (A.) Chrestomathie Française ou Choix de Morceaux tirés des meilleurs Écrivains Français. 11th Edition. 358 pp., cloth. 3*s.* 6*d.*

Roussy. Cours de Versions. Pieces for Translation into French. With Notes. Crown 8vo. 2*s.* 6*d.*

Williams (T. S.) and J. Lafont. French Commercial Correspondence. A Collection of Modern Mercantile Letters in French and English, with their translation on opposite pages. 2nd Edition. 12mo, cloth. 4*s.* 6*d.*
For a German Version of the same Letters, vide p. 4.

Fleury's Histoire de France, racontée à la Jeunesse, with Grammatical Notes, by Auguste Beljame, Bachelier-ès-lettres. 3rd Edition. 12mo, cloth boards. 3*s.* 6*d.*

Mandrou (A.) French Poetry for English Schools. Album Poétique de la Jeunesse. By A. Mandrou, M.A. de l'Académie de Paris. 2nd Edition. 12mo, cloth. 2*s.*

German.

Schlutter's German Class Book. A Course of Instruction based on Becker's System, and so arranged as to exhibit the Self-development of the Language, and its Affinities with the English. By Fr. Schlutter, Royal Military Academy, Woolwich. 5th Edition. 12mo, cloth. (Key, 5*s.*) 5*s.*

Möller (A.) A German Reading Book. A Companion to Schlutter's German Class Book. With a complete Vocabulary.
150 pp. 12mo, cloth.   2s.

Ravensberg (A. v.) Practical Grammar of the German Language.
Conversational Exercises, Dialogues and Idiomatic Ex-
pressions. 3rd Edition. Cloth. (Key, 2s.)   5s.

—— **English into German.** A Selection of Anecdotes,
Stories, &c., with Notes for Translation. Cloth. (Key,
5s.)   4s. 6d.

—— **German Reader,** Prose and Poetry, with copious Notes
for Beginners. 2nd Edition. Crown 8vo, cloth.   3s.

Weisse's Complete Practical Grammar of the German Language,
with Exercises in Conversations, Letters, Poems and
Treatises, &c.   4th Edition, very much enlarged and
improved. 12mo, cloth.   6s.

—— **New Conversational Exercises in German Composition,**
with complete Rules and Directions, with full Refer-
ences to his German Grammar. 2nd Edition. 12mo,
cloth. (Key, 5s.)   3s. 6d.

Wittich's German Tales for Beginners, arranged in Progressive
Order. 26th Edition. Crown 8vo, cloth.   4s.

—— **German for Beginners,** or Progressive German Exercises. 8th Edition. 12mo, cloth. (Key, 5s.)   4s.

—— **German Grammar.** 10th Edition. 12mo, cloth. 4s. 6d.

Hein. German Examination Papers. Comprising a complete
Set of German Papers set at the Local Examinations in
the four Universities of Scotland. By G. Hein, Aberdeen
Grammar School. Crown 8vo, cloth.   2s. 6d.

Schinzel (E.) Child's First German Course; also, A Complete
Treatise on German Pronunciation and Reading. Crown
8vo, cloth.   2s. 6d.

—— **German Preparatory Course.** 12mo, cloth.   2s. 6d.

—— **Method of Learning German.** (A Sequel to the Preparatory Course.) 12mo, cloth.   3s. 6d.

Apel's Short and Practical German Grammar for Beginners, with
copious Examples and Exercises. 3rd Edition. 12mo,
cloth.   2s. 6d.

Sonnenschein and Stallybrass. German for the English. Part I.
First Reading Book. Easy Poems with interlinear Trans-
lations, and illustrated by Notes and Tables, chiefly
Etymological. 4th Edition. 12mo, cloth.   4s. 6d.

Williams (T. S.) Modern German and English Conversations and Elementary Phrases, the German revised and corrected by A. Kokemueller. 21st enlarged and improved Edition. 12mo, cloth. 3s. 6d.

————— and O. Cruse. German and English Commercial Correspondence. A Collection of Modern Mercantile Letters in German and English, with their Translation on opposite pages. 2nd Edition. 12mo, cloth. 4s. 6d.

For a French Version of the same Letters, vide p. 2.

Apel (H.) German Prose Stories for Beginners (including Lessing's Prose Fables), with an interlinear Translation in the natural order of Construction. 12mo, cloth. 2s. 6d.

————— German Prose. A Collection of the best Specimens of German Prose, chiefly from Modern Authors. 500 pp. Crown 8vo, cloth. 3s.

German Classics for English Students. With Notes and Vocabulary. Crown 8vo, cloth.

Schiller's Lied von der Glocke (the Song of the Bell), and other Poems and Ballads. By M. Förster. 2s.

————— Maria Stuart. By M. Förster. 2s. 6d.

————— Minor Poems and Ballads. By Arthur P. Vernon. 2s.

Goethe's Iphigenie auf Tauris. By H. Attwell. 2s.

————— Hermann und Dorothea. By M. Förster. 2s. 6d.

————— Egmont. By H. Apel. 2s. 6d.

Lessing's Emilia Galotti. By G. Hein. 2s.

————— Minna von Barnhelm. By J. A. F. Schmidt. 2s. 6d.

Chamisso's Peter Schlemihl. By M. Förster. 2s.

Andersen's Bilderbuch ohne Bilder. By Alphons Beck. 2s.

Nieritz. Die Waise, a German Tale. By E. C. Otte. 2s. 6d.

Hauff's Maerchen. A Selection. By A. Hoare. 3s. 6d.

Carové (J. W.) Maehrchen ohne Ende (The Story without an End). 12mo, cloth. 2s.

Fouque's Undine, Sintram, Aslauga's Ritter, die beiden Hauptleute. 4 vols. in 1. 8vo, cloth. 7s. 6d.

Undine. 1s. 6d.; cloth, 2s.   Aslauga. 1s. 6d.; cloth, 2s.
Sintram. 2s. 6d.; cloth, 3s.   Hauptleute. 1s. 6d.; cloth, 2s.

Latin and Greek.

Cæsar de Bello Gallico. Lib. I. Edited, with Introduction, Notes and Maps, by Alexander M. Bell, M.A., Ball. Coll. Oxon. Crown 8vo, cloth. 2s. 6d.

Euripides' Medea. The Greek Text, with Introduction and Explanatory Notes for Schools, by J. H. Hogan. 8vo, cloth. 3s. 6d.

—— **Ion.** Greek Text, with Notes for Beginners, Introduction and Questions for Examination, by Dr. Charles Badham, D.D. 2nd Edition. 8vo. 3s. 6d.

Æschylus. Agamemnon. Revised Greek Text, with literal line-for-line Translation on opposite pages, by John F. Davies, B.A. 8vo, cloth. 3s.

Platonis Philebus. With Introduction and Notes by Dr. C. Badham. 2nd Edition, considerably augmented. 8vo, cloth. 4s.

—— **Euthydemus et Laches.** With Critical Notes and an Epistola critica to the Senate of the Leyden University, by Dr. Ch. Badham, D.D. 8vo, cloth. 4s.

—— **Symposium, and Letter to the Master of Trinity, "De Platonis Legibus,"**—Platonis Convivium, cum Epistola ad Thompsonum edidit Carolus Badham. 8vo, cloth. 4s.

Sophocles. Electra. The Greek Text critically revised, with the aid of MSS. newly collated and explained. By Rev. H. F. M. Blaydes, M.A., formerly Student of Christ Church, Oxford. 8vo, cloth. 6s.

—— **Philoctetes.** Edited by the same. 8vo, cloth. 6s.

—— **Trachiniæ.** Edited by the same. 8vo, cloth. 6s.

—— **Ajax.** Edited by the same. 8vo, cloth. 6s.

Dr. D. Zompolides. A Course of Modern Greek, or the Greek Language of the Present Day. I. The Elementary Method. Crown 8vo. 5s.

Kiepert's New Atlas Antiquus. Maps of the Ancient World, for Schools and Colleges. 6th Edition. With a complete Geographical Index. Folio, boards. 7s. 6d.

Kampen. 15 Maps to illustrate Cæsar's De Bello Gallico. 15 coloured Maps. 4to, cloth. 3s. 6d.

Italian.

Volpe (Cav. G.) Eton Italian Grammar, for the use of Eton College. Including Exercises and Examples. New Edition. Crown 8vo, cloth. 4s. 6d.

——— Key to the Exercises. 1s.

Rossetti. Exercises for securing Idiomatic Italian by means of Literal Translations from the English, by Maria F. Rossetti. 12mo, cloth. 3s. 6d.

——— Aneddoti Italiani. One Hundred Italian Anecdotes, selected from "Il Compagno del Passeggio." Being also a Key to Rossetti's Exercises. 12mo, cloth. 2s. 6d.

Venosta (F.) Raccolta di Poesie tratti dai piu celebri autori antichi e moderni. Crown 8vo, cloth. 5s.

Christison (G.) Racconti Istorici e Novelle Morali. Edited for the use of Italian Students. 12th Edition. 18mo, cloth. 1s. 6d.

Danish—Dutch.

Bojesen (Mad. Marie) The Danish Speaker. Pronunciation of the Danish Language, Vocabulary, Dialogues and Idioms for the use of Students and Travellers in Denmark and Norway. 12mo, cloth. 4s.

Williams and Ludolph. Dutch and English Dialogues, and Elementary Phrases. 12mo. 2s. 6d.

Wall Maps.

Sydow's Wall Maps of Physical Geography for School-rooms, representing the purely physical proportions of the Globe, drawn in a bold manner. An English Edition, the Originals with English Names and Explanations. Mounted on canvas, with rollers :

1. The World. 2. Europe. 3. Asia. 4. Africa. 5. America (North and South). 6. Australia and Australasia.
Each 10s.

——— Handbook to the Series of Large Physical Maps for School Instruction, edited by J. Tilleard. 8vo. 1s.

Miscellaneous.

De Rheims (H.). Practical Lines in Geometrical Drawing, containing the Use of Mathematical Instruments and the Construction of Scales, the Elements of Practical and Descriptive Geometry, Orthographic and Horizontal Projections, Isometrical Drawing and Perspective. Illustrated with 300 Diagrams, and giving (by analogy) the solution of every Question proposed at the Competitive Examinations for the Army. 8vo, cloth.   9s.

Fyfe (W. T.) First Lessons in Rhetoric. With Exercises. By W. T. Fyfe, M.A., Senior English Master, High School for Girls, Aberdeen. 12mo, sewed.   1s.

Fuerst's Hebrew Lexicon, by Davidson. A Hebrew and Chaldee Lexicon to the Old Testament, by Dr. Julius Fuerst. 5th Edition, improved and enlarged, containing a Grammatical and Analytical Appendix. Translated by Rev. Dr. Samuel Davidson. 1600 pp., royal 8vo, cloth. 21s.

Strack (W.) Hebrew Grammar. With Exercises, Paradigms, Chrestomathy and Glossary. By Professor H. Strack, D.D., of Berlin. Crown 8vo, cloth.   4s. 6d.

Hebrew Texts. Large type. 16mo, cloth.   Each 1s. Genesis. 1s. Psalms. 1s. Job. 1s. Isaiah. 1s.

Turpie (Rev. Dr.) Manual of the Chaldee Language: containing Grammar of the Biblical Chaldee and of the Targums, and a Chrestomathy, consisting of Selections from the Targums, with a Vocabulary adapted to the Chrestomathy. 1879. Square 8vo, cloth.   7s.

Socin (A.) Arabic Grammar. Paradigms, Literature, Chrestomathy and Glossary. By Dr. A. Socin, Professor Tübingen. Crown 8vo, cloth.   7s. 6d.

Bopp's Comparative Grammar of the Sanscrit, Zend, Greek, Latin, Lithuanian, Gothic, German and Slavonic Languages. Translated by E. B. Eastwick. 4th Edition. 3 vols. 8vo, cloth.   31s. 6d.

Williams and Simmonds. English Commercial Correspondence. A Collection of Modern Mercantile Letters. By T. S. Williams and P. L. Simmonds. 12mo, cloth.   4s.

Williams (T. S.) Modern German and English Conversations and Elementary Phrases, the German revised and corrected by A. Kokemueller. 21st enlarged and improved Edition. 12mo. cloth   3s 6d

Williams (T. S.) and C. Cruse. German and English Commercial Correspondence. A Collection of Modern Mercantile Letters in German and English, with their Translation on opposite pages. 2nd Edition. 12mo. cloth   4s 6d

Apel (H.) German Prose Stories for Beginners (including Lessing's Prose Fables), with an interlinear Translation in the natural order of Construction. 2nd Edition. 12mo. cloth   2s 6d

———— German Prose. A Collection of the best Specimens of German Prose, chiefly from Modern Authors. A Handbook for Schools and Families. 500 pp. Crown 8vo. cloth   3s

German Classics for English Schools, with Notes and Vocabulary. Crown 8vo. cloth.

Schiller's Lied von der Glocke (The Song of the Bell), and other Poems and Ballads, by M. Förster   2s

———— Minor Poems. By Arthur P. Vernon   2s

———— Maria Stuart, by Moritz Förster   2s 6d

Goethe's Hermann und Dorothea, by M. Förster   2s 6d

———— Iphigenie auf Tauris. With Notes by H. Attwell.   2s

———— Egmont. By H. Apel   2s 6d

Lessing's Minna von Barnhelm, by Schmidt   2s 6d

———— Emilia Galotti. By G. Hein   2s

Chamisso's Peter Schlemihl, by M. Förster   2s

Andersen (H. C.) Bilderbuch ohne Bilder, by Beck   2s

Nieritz. Die Waise, a Tale, by Otte   2s

Hauff's Mærchen. A Selection, by A. Hoare   3s 6d

———

Carové (J. W.) Mæhrchen ohne Ende (The Story without an End). 12mo. cloth   2s

Fouque's Undine, Sintram, Aslauga's Ritter, die beiden Hauptleute. 4 vols. in 1. 8vo. cloth   7s 6d

Undine. 1s 6d; cloth, 2s. Aslauga. 1s 6d; cloth, 2s
Sintram. 2s 6d; cloth, 3s. Hauptleute. 1s 6d; cloth, 2s

Latin, Greek, etc.

Cæsar de Bello Gallico. Lib. I. Edited with Introduction, Notes and Maps, by ALEXANDER M. BELL, M.A. Ball. Coll., Oxon. Crown 8vo. cloth   2s 6d

Euripides' Medea. The Greek Text, with Introduction and Explanatory Notes for Schools, by J. H. Hogan. 8vo. cloth   3s 6d

——— **Ion.** Greek Text, with Notes for Beginners, Introduction and Questions for Examination, by the Rev. Charles Badham, D.D. 2nd Edition. 8vo. 3s 6d

Æschylus. Agamemnon. Revised Greek Text, with literal line-for-line Translation on opposite pages, by John F. Davies, B.A. 8vo. cloth   3s

Platonis Philebus. With Introduction and Notes by Dr. C. Badham. 2nd Edition, considerably augmented. 8vo. cloth   4s

——— **Euthydemus et Laches.** With Critical Notes, by the Rev. Ch. Badham, D.D. 8vo. cloth   4s

——— **Convivium,** cum Epistola ad Thompsonum, " De Platonis Legibus," edidit C. Badham. 8vo. cloth 4s

Dr. D. Zompolides. A Course of Modern Greek, or the Greek Language of the Present Day. I. The Elementary Method. Crown 8vo.   5s

---

Kiepert New Atlas Antiquus. Maps of the Ancient World, for Schools and Colleges. 6th Edition. With a complete Geographical Index. Folio, boards   7s 6d

Kampen. 15 Maps to illustrate Cæsar's De Bello Gallico. 15 coloured Maps. 4to. cloth   3s 6d

---

Italian.

Volpe (Cav. G.) Eton Italian Grammar, for the use of Eton College. Including Exercises and Examples. New Edition. Crown 8vo. cloth (Key, 1s)   4s 6d

Racconti Istorici e Novelle Morali. Edited, for the use of Italian Students, by G. Christison. 12th Edition. 18mo. cloth   1s 6d

Rossetti. Exercises for securing Idiomatic Italian, by means of Literal Translations from the English by Maria F. Rossetti. 12mo. cloth   3s 6d

——— **Aneddoti Italiani.** One Hundred Italian Anecdotes, selected from " Il Compagno del Passeggio."

www.ingramcontent.com/pod-product-compliance
Lightning Source LLC
Chambersburg PA
CBHW021420090426
42742CB00009B/1197